Modern Food Processing Biotechnology

Modern Food Processing Biotechnology

Ranvijay Singh
Editor

KOROS PRESS LIMITED
London, UK

Modern Food Processing Biotechnology

© 2012

Printed in 2017 for Sale in the Indian Subcontinent

Published by
Koros Press Limited
3 The Pines, Rubery B45 9FF, Rednal,
Birmingham, United Kingdom

Tel.: +44-7826-930152
Email: info@korospress.com
www.korospress.com

ISBN: 978-1-78163-183-6

Editor: Ranvijay Singh

Printed in U.K.

British Library Cataloguing in Publication Data
A CIP record for this book is available from the British Library

10 9 8 7 6 5 4 3 2 1

No part of this publication maybe reproduced, stored in a retrieval system or transmitted in any form or by any means, electronic, mechanical, photocopying, recording, scanning or otherwise without prior written permission of the publisher.

Reasonable efforts have been made to publish reliable data and information, but the authors, editors, and the publisher cannot assume responsibility for the legality of all materials or the consequences of their use. The authors, editors, and the publisher have attempted to trace the copyright holders of all material in this publication and express regret to copyright holders if permission to publish has not been obtained. If any copyright material has not been acknowledged, let us know so we may rectify in any future reprint.

Exclusively distributed by CBS Publishers & Distributors Pvt. Ltd.
Sales & Distribution Rights only for India, Pakistan, Bangladesh, Sri Lanka, Nepal and Bhutan. This book is not to be sold outside these territories.

Contents

Preface   *vii*

1. **Food Processing**   1
   Food Sources • Food Chain • Trophic Level • Ecological Efficiency • Pyramids • Animal Source Foods • Production • Cooking • Cooking Methods • Barbecuing – Grilling – Rotisserie – Searing • Baking Blind - Broiling - Flashbaking • Boiling Stages • Frying • Smoking • Food Safety • Regulatory Agencies • Manufacturing Control • Consumer Labelling • Raw Food Preparation • Food Manufacture • Production Processes • Commercial Trade

2. **Enzymes in Food Processing**   59
   Enzymes in Food Processing: A Condensed Overview on Strategies for Better Biocatalysts • Relevant Enzymes: Tapping for Improved Biocatalysts • Improving Biocatalysts: Beyond Screening • Immobilization • Conclusions and Future Perspectives

3. **Application of Modern Biotechnology in Food Processing**   83
   Industrial Biotechnology • Environmental Biotechnology • Biotechnology for the 21st Century • Consumer and Food Industry Perspectives • National and International Biotechnology Policy • The 6 Top Trends in Food Processing • Modern Biotechnology in Food: Modern biotechnology and Food Quality • Beverages Take Two Paths • Technological Trends and Needs • Nutritional Quality • Energy Provided by Macronutrients • Energy Density Concept • Macronutrients • Fat • Carbohydrates • Nutritionally Improved Food Feeds • Nutritional Assessment Process for Biotechnology • The Concept of Substantial Equivalence, its Historical • Development of a Safety Assessment Framework for GM Foods

4. **Enzymology for the Food Science**   135
Immobilized Enzymes • Various Methods Used for Immobilization of Enzymes • Entrapment of Enzymes and Microbial Cells • Proteases in Food Processing • Proteases in Detergent Formulation • Medical Applications of Enzymes • Glucose Oxidase Electrode

5. **Microbiology used in Wine Production**   167
Wine is Fermented Grape Juice • How is Wine Made? • Crushing and Primary Fermentation • Preservatives • Bottling Wine • Cellaring Wine • Pectic Enzymes on Clarification of Wines • Pectinase • Sources and Production • The Effect of Pectic Enzymes in Wine Making • Vinegar Fermentation and Production • Production • Agricultural and Horticultural Uses

6. **Calcium and Sodium in Food Nutrition**   199
Calcium in Nutrition • Notable Characteristics • Nucleosynthesis • Geochemical Cycling • Nutrition • Calcium in Biology • Calcium Metabolism • Calcium Location and Quantity • Effector Organs • Interaction with Other Chemicals • Research into Cancer Prevention • Dietary Calcium Supplements • Prevention of Fractures due to Osteoporosis • Hazards and Toxicity • Sodium in Nutrition • Commercial Production • Applications • Biological Role • Dietary Uses • Salt • Forms of Salt • Health Effects • Sodium-Metabolism • Tissue Distribution • Precautions

7. **Homocysteine an Amino Acid**   234
Methionine Salvage • Influence, Proposed and Verified of Homocysteine on Human Health • Cofactors or Coenzymes: Nature's Special Reagents • Inorganic • Organic • Abiogenesis • Conceptual History • Models to Explain Homochirality • Other Models • Extraterrestrial Amino Acids • PAH World Hypothesis • Coenzymes of Oxidation Reduction Reactions • Population Nutrition Health Promotion and Government Policy

Bibliography   273

Index   277

Preface

Survey research over the past decade shows that biotechnology is not likely to become an important issue for most American consumers. Consumers find biotechnology acceptable when they believe it offers benefits and it is safe. Surveys have consistently found that a majority of American consumers are willing to buy insect-protected food crops developed through biotechnology that use fewer chemical pesticides, as well as more nutritious foods. American consumers also appreciate the role that biotechnology can play in feeding the world. Research shows that European consumers are much less supportive of all biotechnology applications. Surveys since 1992 show that relatively few U.S. consumers have heard or read much about biotechnology. News about the cloned sheep pushed awareness to 50 percent in March 1997. Surveys in the first three months of 2000 show that awareness has fallen back to just over one-third in the United States. Such trends reflect the fact that most people get their information about biotechnology from the media. Unfortunately, many consumers also do not understand some fundamental principles of biology. European consumer awareness is somewhat higher, but knowledge is still low.

Media coverage in the United States has generally been balanced (which helps account for our relatively high levels of acceptance). This is in sharp contrast to the European media, which have played upon fear of the unknown. The European media have also tended to accept opponents' claims without question. Another issue is that many people no longer have a connection to agriculture. In fact, research has shown that many consumers are unaware that all foods are derived from plants or animals that already have been genetically modified through traditional (but imprecise) breeding methods. American consumers look to health professionals and scientific experts for credible information, but place relatively little trust in the activists who oppose biotechnology. Research shows that acceptance increases significantly when American consumers learn that organizations such as the National Academy of Sciences and the U.S. Food and Drug Administration have determined that biotech-derived foods are safe.

In contrast, European consumers express the most trust in those groups that oppose biotechnology. They have much less confidence in government, industry, or even scientists. American culture is more supportive and rewarding of new technology.

Europeans tend to view food differently from U.S. consumers. In fact, some Europeans reject all American food products. Europeans also want to protect their small farms to maintain open space and rural employment. Such forces underlie much of the European anxiety about agricultural biotech - especially since it is seen as an "American invention." Most of the industry leaders interviewed are quite enthusiastic about the benefits of biotechnology — especially in terms of increased food availability, enhanced nutrition, and environmental protection. Most feel that biotechnology has already provided benefits to consumers. Almost all recognize that foods developed through biotechnology have already been part of consumers' everyday diet. They clearly do not agree with most of the opponents' claims and tend to have almost no trust in such groups. Their main concerns involve lack of consumer acceptance — not the safety of the foods. They express high levels of confidence in the science and the regulatory process. In fact, almost none feel that biotechnology should not be used because of uncertain, potential risks. Most food industry leaders do not feel it is necessary to have special labels on biotech-derived foods. They express concerns that such labels would be perceived as a warning by consumers. They also worried that the need to segregate commodities would pose financial and logistical burdens on everyone in the system - including consumers. Food industry leaders recognize a major need to educate the public about biotechnology. They look to third parties, such as university and government scientists to provide such leadership.

The book has been written keeping in mind for the graduate and post graduate level bio-sciences and interdisciplinary courses.

—Ranvijay Singh

Chapter 1
Food Processing

Food is any substance or materials that is consumed to provide nutritional support for the body or for pleasure. It is usually of plant or animal origin, and contains essential nutrients, such as carbohydrates, fats, proteins, vitamins, or minerals. It is ingested and assimilated by an organism to produce energy, stimulate growth, and maintain life.

Historically, people obtained food from hunting and gathering, farming, ranching, and fishing, known as agriculture. Today, most of the food energy consumed by the world population is supplied by the food industry operated by multinational corporations using intensive farming and industrial agriculture methods.

Food safety and food security are monitored by agencies such as the International Association for Food Protection, World Resources Institute, World Food Programme, Food and Agriculture Organization, and International Food Information Council. They address issues such as sustainability, biological diversity, climate change, nutritional economics, population growth, water supply, and access to food.

The right to food is a human right derived from the International Covenant on Economic, Social and Cultural Rights (ICESCR), recognizing the "right to an adequate standard of living, including adequate food", as well as the "fundamental right to be free from hunger".

Food Sources

Almost all foods are of plant or animal origin. Cereal grain is a staple food that provides more food energy worldwide than any other type of crop. Maize, wheat, and rice together account for 87% of all grain production worldwide.

Other foods not from animal or plant sources include various edible fungi, especially mushrooms. Fungi and ambient bacteria are used in the preparation of fermented and pickled foods such as leavened bread, alcoholic drinks, cheese, pickles, kombucha, and yogurt. Another example is blue-green algae such as Spirulina. Inorganic substances such as baking soda and cream of tartar are also used to chemically alter an ingredient.

Plants

Many plants or plant parts are eaten as food. There are around 2,000 plant species which are cultivated for food, and many have several distinct cultivars.

Seeds of plants are a good source of food for animals, including humans, because they contain the nutrients necessary for the plant's initial growth, including many healthy fats, such as Omega fats. In fact, the majority of food consumed by human beings are seed-based foods. Edible seeds include cereals (such as maize, wheat, and rice), legumes (such as beans, peas, and lentils), and nuts. Oilseeds are often pressed to produce rich oils, such as sunflower, flaxseed, rapeseed (including canola oil), and sesame. Seeds are typically high in unsaturated fats and, in moderation, are considered a health food, although not all seeds are edible. Large seeds, such as those from a lemon, pose a choking hazard, whereas seeds from apples and cherries contain the poison cyanide.

Fruits are the ripened ovaries of plants, including the seeds within. Many plants have evolved fruits that are attractive as a food source to animals, so that animals will eat the fruits and excrete the seeds some distance away. Fruits, therefore, make up a significant part of the diets of most cultures. Some botanical fruits, such as tomatoes, pumpkins, and eggplants, are eaten as vegetables.

Vegetables are a second type of plant matter that is commonly eaten as food. These include root vegetables (such as potatoes and carrots), leaf vegetables (such as spinach and lettuce), stem vegetables (such as bamboo shoots and asparagus), and inflorescence vegetables (such as globe artichokes and broccoli). Many herbs and spices are highly flavour some vegetables.

Animals

Animals are used as food either directly or indirectly by the products they produce. Meat is an example of a direct product taken from an animal, which comes from either muscle systems or from organs. Food products produced by animals include milk produced by mammary glands, which in many cultures is drunk or processed into dairy products such as cheese or butter. In addition, birds and other animals lay eggs, which are often eaten, and bees produce honey, a reduced nectar from flowers, which is a popular sweetener in many cultures. Some cultures consume blood, sometimes in the form of blood sausage, as a thickener for sauces, or in a cured, salted form for times of food scarcity, and others use blood in stews such as civet.

Some cultures and people do not consume meat or animal food products for cultural, dietary, health, ethical, or ideological reasons. Vegetarians do not consume meat. Vegans do not consume any foods that are or contain ingredients from an animal source.

Food Chain

Food chains and food webs are representations of the predator-prey relationships between species within an ecosystem or habitat.

Many chain and web models can be applicable depending on habitat or environmental factors. Every known food chain has a base made of autotrophs, organisms able to manufacture their own food (e.g. plants, chemotrophs).

Organisms Represented in Food Chains

In nearly all food chains, solar energy is input into the system as light and heat, utilized by autotrophs (i.e., producers) in a process called photosynthesis. Carbon dioxide is reduced (gains electrons) by being combined with water (a source of hydrogen atoms), producing glucose. Water splitting produces hydrogen, but is a nonspontaneous (endergonic) reaction requiring energy from the sun. Carbon dioxide and water, both stable, oxidized compounds, are low in energy, but glucose, a high-energy compound and good electron donor, is capable of storing the solar energy. This energy is expended for cellular processes, growth, and development. The plant sugars are polymerized for storage as long-chain carbohydrates, including other sugars, starch, and cellulose.

Glucose is also used to make fats and proteins. Proteins can be made using nitrates, sulfates, and phosphates in the soil. When autotrophs are eaten by heterotrophs, i.e., consumers such as animals, the carbohydrates, fats, and proteins contained in them become energy sources for the heterotrophs.

Chemoautotrophy

An important exception is lithotrophy, the utilization of inorganic compounds, especially minerals such as sulphur or iron, for energy. In some lithotrophs, minerals are used simply to power processes for making organic compounds from inorganic carbon sources. In a few food chains, e.g., near hydrothermal vents in the deep sea, autotrophs are able to produce organic compounds without sunlight, through a process similar to photosynthesis called chemosynthesis, using a carbon source such as carbon dioxide and a chemical energy sources such as hydrogen sulfide, H_2S, or molecular hydrogen, H_2.

Unlike water, the hydrogen compounds used in chemosynthesis are high in energy. Other lithotrophs are able to directly utilize inorganic substances, e.g., iron, hydrogen sulfide, elemental sulphur, or thiosulfate, for some or all of their energy needs.

Involvement in the Carbon Cycle

Carbon dioxide is recycled in the carbon cycle as carbohydrates, fats, and proteins are oxidized (burned) to produce carbon dioxide and water. Oxygen released by photosynthesis is utilized in respiration as an electron acceptor to release chemical energy stored in organic compounds. Dead organisms are consumed by detritivores, scavengers, and decomposers, including fungi and insects, thus returning nutrients to the soil.

Food Web

Food chains are overly simplistic as representatives of the relationships of living organisms in nature. Most consumers feed on multiple species and in turn, are fed upon by multiple other species.

For a snake, the prey might be a mouse, a lizard, or a frog, and the predator might be a bird of prey or a badger. The relations of detritivores and parasites are seldom adequately characterized in such chains as well.

A food web is a series of related food chains displaying the movement of energy and matter through an ecosystem. The food web is divided into two broad categories: the grazing web, beginning with autotrophs, and the detrital web, beginning with organic debris. There are many food chains contained in these food webs.

In a grazing web, energy and nutrients move from plants to the herbivores consuming them to the carnivores or omnivores preying upon the herbivores. In a detrital web, plant and animal matter is broken down by decomposers, e.g., bacteria and fungi, and moves to detritivores and then carnivores.

There are often relationships between the detrital web and the grazing web. Mushrooms produced by decomposers in the detrital web become a food source for deer, squirrels, and mice in the grazing web.

Earthworms eaten by robins are detritivores consuming decaying leaves.

Flow of Food Chains

Food energy flows from one organism to the next and to the next and so on, with some energy being lost at each level. Organisms in a food chain are grouped into trophic levels, based on how many links they are removed from the primary producers. In trophic levels there may be one species or a group of species with the same predators and prey.

Autotrophs such as plants or phytoplankton are in the first trophic level; they are at the base of the food chain. Herbivores (primary consumers) are in the second trophic level. Carnivores (secondary consumers) are in the third. Omnivores are found in the second and third levels. Predators preying upon other predators are tertiary consumers or secondary carnivores, and they are found in the fourth trophic level.

Food chain length is another way of describing food webs as a measure of the number of species encountered as energy or nutrients move from the plants to top predators. There are different ways of calculating food chain length depending on what parameters of the food web dynamic are being considered: connectance, energy, or interaction. In a simple predator-prey example, a deer is one step removed from the plants it eats (chain length = 1) and a wolf that eats the deer is two steps removed (chain length = 2). The relative amount or strength of influence that these parameters have on the food web address questions about:

- the identity or existence of a few dominant species (called strong interactors or keystone species)
- the total number of species and food-chain length (including many weak interactors) and
- how community structure, function and stability is determined.

Trophic Level

The trophic level of an organism is the position it occupies on the food web. The word trophic derives from the Greek τροφη (trophe) referring to food or feeding. A food chain represents a succession of organisms that eat another organism and are, in turn, eaten themselves. The number of steps an organism is from the start of the chain is a measure of its trophic level. Food chains start at trophic level 1 with primary producers such as plants, move to herbivores at level 2, predators at level 3 and typically finish with carnivores or apex predators at level 4 or 5. The path along the chain can form a one-way flow, or a food "web." Ecological communities with higher biodiversity form more complex trophic paths.

The plants in this image, and the algae and phytoplankton in the lake, are primary producers. They take nutrients from the soil or the water, and manufacture their own food by photosynthesis, using energy from the sun. The three basic ways organisms get food are as producers, consumers and decomposers.

- Producers (autotrophs) are typically plants or algae. Plants and algae do not usually eat other organisms, but pull nutrients from the soil or the ocean and manufacture their own food using photosynthesis. For this reason, they are called primary producers. In this way, it is energy from the sun that usually powers the base of the food chain. An exception occurs in deep-sea hydrothermal ecosystems, where there is no sunlight. Here primary producers manufacture food through a process called chemosynthesis.
- Consumers (heterotrophs) are animals which cannot manufacture their own food and need to consume other organisms. Animal that eat primary producers (like plants) are called herbivores. Animals that eat other animals are called carnivores, and animals that eat both plant and other animals are called omnivores.
- Decomposers (detritivores) break down dead plant and animal material and wastes and release it again as energy and nutrients into the ecosystem for recycling. Decomposers, such as bacteria and fungi (mushrooms), feed on waste and dead matter, converting it into inorganic chemicals that can be recycled as mineral nutrients for plants to use again.

Trophic levels can be represented by numbers, starting at level 1 with plants. Further trophic levels are numbered subsequently according to how far the organism is along the food chain.

- Level 1: Plants and algae make their own food and are called primary producers.
- Level 2: Herbivores eat plants and are called primary consumers.
- Level 3: Carnivores which eat herbivores are called secondary consumers.

- Level 4: Carnivores which eat other carnivores are called tertiary consumers.
- Level 5: Apex predators which have no predators are at the top of the food chain.

In real world ecosystems, there is more than one food chain for most organism, since most organisms eat more than one kind of food or are eaten by more than one type of predator. A diagram which sets out the intricate network of intersecting and overlapping food chains for an ecosystem is called its food web. Decomposers are often left off food webs, but if included, they mark the end of a food chain. Thus food chains start with primary producers and end with decay and decomposers. Since decomposers recycle nutrients, leaving them so they can be reused by primary producers, they are sometimes regarded as occupying their own trophic level.

Biomass Transfer Efficiency

Generally, each trophic level relates to the one below it by absorbing some of the energy it consumes, and in this way can be regarded as resting on, or supported by the next lower trophic level. Food chains can be diagrammed to illustrate the amount of energy that moves from one feeding level to the next in a food chain. This is called an energy pyramid. The energy transferred between levels can also be thought of as approximating to a transfer in biomass, so energy pyramids can also be viewed as biomass pyramids, picturing the amount of biomass that results at higher levels from biomass consumed at lower levels.

The efficiency with which energy or biomass is transferred from one trophic level to the next is called the ecological efficiency. Consumers at each level convert on average only about 10 percent of the chemical energy in their food to their own organic tissue. For this reason, food chains rarely extend for more than 5 or 6 levels. At the lowest trophic level (the bottom of the food chain), plants convert about one percent of the sunlight they receive into chemical energy. It follows from this that the total energy originally present in the incident sunlight that is finally embodied in a tertiary consumer is about 0.001 %

Fractional Trophic Levels

Food webs largely define ecosystems, and the trophic levels define the position of organisms within the webs. But these trophic levels are not always simple integers, because organisms often feed at more than one trophic level. For example, some carnivores also eat plants, and some plants are carnivores. A large carnivore may eat both smaller carnivores and herbivores; the bobcat eats rabbits, but the mountain lion eats both bobcats and rabbits. Animals can also eat each other; the bullfrog eats crayfish and crayfish eat young bullfrogs. The feeding habits of a juvenile animal, and consequently its trophic level, can change as it grows up. The fisheries scientist Daniel Pauly sets the values of trophic levels to one in plants and detritus, two in herbivores and detritivores (primary consumers), three in secondary consumers, and so on. The definition of the trophic level, TL, for any consumer species i is:

$$TL_i = 1 + \sum_j \left(TL_j . DC_{ij} \right)$$

where TL_i is the fractional trophic level of the prey j, and DC_{ij} represents the fraction of j in the diet of i. In the case of marine ecosystems, the trophic level of most fish and other marine consumers takes value between 2.0 and 5.0. The upper value, 5.0, is unusual, even for large fish, though it occurs in apex predators of marine mammals, such as polar bears and killer whales.

Mean trophic level

In fisheries, the mean trophic level for the fisheries catch across an entire area or ecosystem is calculated for year y as:

$$TL_y = \frac{\sum_i \left(TL_i . Y_{iy} \right)}{\sum_i Y_{iy}}$$

where Y_{iy} is the catch of the species or group i in year y, and TL_i is the fractional trophic level for species i as defined above.

It was once believed that fish at higher trophic levels usually have a higher economic value; resulting in overfishing at the higher trophic levels. Earlier reports found precipitous declines in mean trophic level of fisheries catch, in a process known as fishing down the food web. However, more recent work finds no relation between economic value and trophic level; and that mean trophic levels in catches, surveys and stock assessments have not in fact declined, suggesting that fishing down the food web is not a global phenomenon.

FiB Index

Since biomass transfer efficiencies are only about 10 percent, it follows that the rate of biological production is much greater at lower trophic levels than it is at higher levels. Fisheries catches, at least to begin with, will tend to increase as the trophic level declines. At this point the fisheries will target species lower in the food web. In 2000, this led Pauly and others to construct a "Fisheries in Balance" index, usually called the FiB index. The FiB index is defined, for any year y, by

$$FiBy = \log ((Y_y/(TE)^{TLy})/(Y_0/(TE)^{TL0}))$$

where Y_y is the catch at year y, TL_y is the mean trophic level of the catch at year y, Y_0 is the catch and TL_0 the mean trophic level of the catch at the start of the series being analyzed, and TE is the transfer efficiency of biomass or energy between trophic levels.

The FiB index is stable (zero) over periods of time when changes in trophic levels are matched by appropriate changes in the catch in the opposite direction. The index increases if catches increase for any reason, e.g. higher fish biomass, or geographic expansion Such decreases explain the "backward-bending" plots of trophic level versus catch originally observed by Pauly and others in 1998.

Entropic Losses in the Chain

It is the case that the biomass of each trophic level decreases from the base of the chain to the top. This is because energy is lost to the environment with each transfer as entropy increases. About eighty to ninety percent of the energy is expended for the organism's life processes or is lost as heat or waste. Only about ten to twenty percent of the organism's energy is generally passed to the next organism. The amount can be less than one percent in animals consuming less digestible plants, and it can be as high as forty percent in zooplankton consuming phytoplankton. Graphic representations of the biomass or productivity at each tropic level are called ecological pyramids or trophic pyramids. The transfer of energy from primary producers to top consumers can also be characterized by energy flow diagrams.

Ecological Efficiency

Ecological efficiency describes the efficiency with which energy is transferred from one trophic level to the next. It is determined by a combination of efficiencies relating to organismic resource acquisition and assimilation in an ecosystem.

Energy Transfer

Primary production occurs in autotrophic organisms of an ecosystem. Photoautotrophs such as vascular plants and algae convert energy from the sun into energy stored as carbon compounds.

Photosynthesis is carried out in the chlorophyll of green plants. The energy converted through photosynthesis is carried through the trophic levels of an ecosystem as organisms consume members of lower trophic levels.

Primary production can be broken down into gross and net primary production. Gross primary production is a measure of the energy that a photoautotroph harvests from the sun. Take, for example, a blade of grass that takes in x Joules of energy from the sun. The fraction of that energy that is converted into glucose reflects the gross productivity of the blade of grass. The energy remaining after respiration is considered the net primary production. In general, gross production refers to the energy contained within an organism before respiration and net production the energy after respiration. The terms can be used to describe energy transfer in both autotrophs and heterotrophs.

Energy transfer between trophic levels is generally inefficient, such that net production at one trophic level is generally only 10% of the net production at the preceding trophic level. Due to non-predatory death, egestion, and respiration, a significant amount of energy is lost to the environment instead of being absorbed for production by consumers. The figure approximates the fraction of energy available after each stage of energy loss in a typical ecosystem, although these fractions vary greatly from ecosystem to ecosystem and from trophic level to trophic level.

Example: Assume 500 units of energy are produced by trophic level 1. One half of that is lost to non-predatory death, while the other half (250 units) is ingested by trophic level 2. One half of the amount ingested is expelled through defecation, leaving the other half (125 units) to be assimilated by the organism. Finally one half of the remaining energy is lost through respiration while the rest (63 units) is used for growth and reproduction. This energy expended for growth and reproduction constitutes to the net production of trophic level 1, which is equal to 500 * 1/2 * 1/2 * 1/2 = 63 units.

Quantifying Ecological Efficiency

Ecological efficiency is a combination of several related efficiencies that describe resource utilization and the extent to which resources are converted into biomass.

- Exploitation efficiency is the amount of food ingested divided by the amount of prey production (I/P_n)

- Assimilation efficiency is the amount of assimilation divided by the amount of food ingestion (A/I)
- Net Production efficiency is the amount of consumer production divided by the amount of assimilation (P_{n+1}/A)
- Gross Production efficiency is the assimilation efficiency multiplied by the net production efficiency, which is equivalent to the amount of consumer production divided by amount of ingestion (P_{n+1}/I)
- Ecological efficiency is the exploitation efficiency multiplied by the assimilation efficiency multiplied by the net production efficiency, which is equivalent to the amount of consumer production divided by the amount of prey production (P_{n+1}/P_n).

Theoretically, it is easy to calculate ecological efficiency using the mathematical relationships above. It is often difficult, however, to obtain accurate measurements of the values involved in the calculation. Assessing ingestion, for example, requires knowledge of the gross amount of food consumed in an ecosystem as well as its caloric content. Such a measurement is rarely better than an educated estimate, particularly with relation to ecosystems that are largely inaccessible to ecologists and tools of measurement. The ecological efficiency of an ecosystem is as a result often no better than an approximation. On the other hand, an approximation may be enough for most ecosystems, where it is important not to get an exact measure of efficiency, but rather a general idea of how energy is moving through its trophic levels.

Applications

In agricultural environments, maximizing energy transfer from producer (food) to consumer (livestock) can yield economic benefits. A sub-field of agricultural science has emerged that explores methods of monitoring and improving ecological and related efficiencies.

In comparing the net efficiency of energy utilization by cattle, breeds historically kept for beef production, such as the Hereford, outperformed those kept for dairy production, such as the Holstein, in converting energy from feed into stored energy as tissue. This is a result of the beef cattle storing more body fat than the dairy cattle, as energy storage as protein was at the same level for both breeds. This implies that cultivation of cattle for slaughter is a more efficient use of feed than is cultivation for milk production.

While it is possible to improve the efficiency of energy use by livestock, it is vital to the world food question to also consider the differences between animal husbandry and plant agriculture. Caloric concentration in fat tissues are higher than in plant tissues, causing high-fat organisms to be most energetically-concentrated; however, the energy required to cultivate feed for livestock is only partially converted into fat cells. The rest of the energy input into cultivating feed is respirated or egested by the livestock and unable to be used by humans. Out of a total of $96.8 * 10^{15}$ BTU of energy used in the US in 1999, 10.5% was used in food production, with the percentage accounting for food from both producer and primary consumer trophic levels. In comparing the cultivation of animals versus plants, there is a clear difference in magnitude of energy efficiency. Edible kilocalories produced from kilocalories of energy required for cultivation are: 18.1% for chicken, 6.7% for grass-fed beef, 5.7% for farmed salmon, and 0.9% for shrimp. In contrast, potatoes yield 123%, corn produce 250%, and soy results in 415% of input calories converted to calories able to be utilized by humans. This disparity in efficiency reflects the reduction in production from moving up trophic levels. Thus, it is more energetically efficient to form a diet from lower trophic levels.

Pyramids

In a pyramid of numbers, the number of consumers at each level decreases significantly, so that a single top consumer, (e.g., a polar bear or a human), will be supported by a million separate producers. There is usually a maximum of four or five links in a food chain, although food chains in aquatic ecosystems are frequently longer than those on land. Eventually, all the energy in a food chain is lost as heat.

Some producers, especially phytoplankton, are able to reproduce quickly enough to support a larger biomass of grazers. This is called an inverted pyramid, caused by a longer lifespan and slower growth rate in the consumers than in the organisms being consumed, with phytoplankton living just a few days, compared to several weeks for the zooplankton eating the phytoplankton and years for fish eating the zooplankton. A pyramid of energy, reflecting the energy or kilojoules in each level, is representative of the true relationships of the phytoplankton, zooplankton, and fish, showing phytoplankton as the largest section, then zooplankton as a smaller section, and fish as the smallest section.

History of Food Webs

Food webs serve as a framework to help ecologists organize the complex network of interactions among species observed in nature. The earliest description of a food chain was given by the medieval

Afro-Arab biologist Al-Jahiz (781-868). Perhaps the earliest graphical depiction of a food web was by Lorenzo Camerano in 1880, followed independently by those of Pierce and colleagues in 1912 and Victor Shelford in 1913. Two food webs about herring were produced by Victor Summerhayes and Charles Elton and Alister Hardy in 1923 and 1924. After Charles Elton's use of food webs in his 1927 synthesis, they became a central concept in the field of ecology. The utilization of the common currency of energy flow along links in a flow was emphasized in Raymond Lindeman's work, initiating the extensive analysis of energy and material flows that are a core activity of ecosystem ecology.

Interest in food webs increased after Robert Paine's experimental and descriptive study of intertidal shores suggesting that food web complexity was key to maintaining species diversity and ecological stability. Many theoretical ecologists, including Sir Robert May and Stuart Pimm, were prompted by this discovery and others to examine the mathematical properties of food webs. According to their analyses, complex food webs should be highly unstable. The apparent paradox between the complexity of food webs observed in nature and the mathematical fragility of food web models is currently an area of intensive study and debate. The paradox may be due partially to conceptual differences between persistence of a food web and equilibrial stability of a food web. Current research points to important roles of non-random structure in the connections within the food web that develop as food webs assemble over long periods of time, of patterns in the strengths of interactions among species within the food web, of variable strengths of species interactions as species abundances change, and of spatial variation in the environment creating food webs of different structures that are connected by movement of individuals and materials, in the creation and persistence of complex food webs.

Animal Source Foods

Animal source foods (ASF) include any food item that comes from an animal source such as meat, milk, fish, eggs, cheese and yogurt. Many individuals do not consume ASF or consume little ASF by either personal choice or necessity as ASF may not be accessible or available to these people.

Nutrition of Animal Source Foods

Aside from performed vitamin A, vitamin B_{12} and vitamin D, all vitamins found in animal source foods may also be found in plant-derived foods. Examples are tofu to replace meat (both contain protein in sufficient amounts), and certain seaweeds and vegetables as respectively kombu and kale to replace dairy foods as milk (both contain calcium in sufficient amounts). Some plant-derived foods are nutrient-denser than their animal-derived counterparts (e.g. tofu).

Most humans eat an omnivorous diet (comprising animal source foods and plant source foods) though some civilisations have eaten only animal foods. Although a healthy diet containing all essential macro and micronutrients may be possible by only consuming a plant based diet (with vitamin B_{12} obtained from supplements if no animal sourced foods are consumed), some populations are unable to consume an adequate quantity or variety of these plant based items to obtain appropriate amounts of nutrients, particularly those that are found in high concentrations in ASF. Frequently, the most vulnerable populations to these micronutrient deficiencies are pregnant women, infants, and children in developing countries. In the 1980s the Nutrition Collaborative Research Support Program (NCRSP) found that six micronutrients were low in the mostly vegetarian diets of children in malnourished areas of Egypt, Mexico, and Kenya. These six micronutrients are vitamin A, vitamin B_{12}, riboflavin, calcium, iron and zinc. ASF are the only food source of Vitamin B_{12}. ASF also provide high biological value protein, energy, fat compared with plant food sources.

Health Impacts of Micronutrient Deficiency

All six micronutrients richly found in ASF, vitamin A, vitamin B_{12}, riboflavin, calcium, iron and zinc play a critical role in the growth and development of children. Inadequate stores of these micronutrients, either resulting from inadequate intake or poor absorption, is associated with poor growth, anemias (iron deficiency anemia and macrocytic anemia), rickets, night blindness, impaired cognitive functioning, neuromuscular deficits, diminished work capacity, psychiatric disorders and death. Some of these affects, such as impaired cognitive development from an iron deficiency, are irreversible.girja

Animal Source Food Supplementation

Micronutrient deficiency is associated in poor early cognitive development. Programs designed to address these micronutrient deficiencies should be targeted to infants, children, and pregnant women. To address these significant mirconutrient deficiencies, some global health researchers and practitioners developed and piloted a snack program in Kenya school children. However, some communities are vegetarians for religious or cultural reasons. Efforts must be made to develop culturally appropriate interventions to address the micronutrient deficiencies in these populations, such as through food fortification.

Animal Source Food Production

According to a 2006 United Nations initiative, the livestock industry sector emerges as one of the top two or three most significant contributors to the most serious environmental problems, at every scale from local to global." As such, using plant-derived foods is always better in the interest of the environment. Despite this, elevation of certain animals are (somewhat) more environmental than others. According to the Farralones Institute, elevation of rabbits, and chicken (on a well-considered approach) can still be done quite environmental. As such, meat and other produce as eggs may still be produced quite environmental (if this is done on an industrial, high-efficiency manner). In addition, the elevation of goats (and their produce as goat milk and meat) can too be done quite environmentally and has been favored by certain environmental activists as Mohandas Gandhi.

Production

Traditionally, food was obtained through agriculture. With increasing concern in agri-business over multinational corporations owning the world food supply through patents on genetically modified food, there has been a growing trend toward sustainable agricultural practices. This approach, partly fueled by consumer demand, encourages biodiversity, local self-reliance and organic farming methods. Major influences on food production include international organizations (e.g. the World Trade Organization and Common Agricultural Policy), national government policy (or law), and war.

In popular culture, the mass production of food, specifically meats such as chicken and beef, has come under fire from various documentaries, most recently Food, Inc, documenting the mass slaughter and poor treatment of animals, often for easier revenues from large corporations. Along with a current trend towards environmentalism, people in Western culture have had an increasing trend towards the use of herbal supplements, foods for a specific group of person (such as dieters, women, or athletes), functional foods (fortified foods, such as omega-3 eggs), and a more ethnically diverse diet.

Cuisine Preparation

Many cultures have a recognizable cuisine, a specific set of cooking traditions using various spices or a combination of flavours unique to that culture, which evolves over time. Other differences include preferences (hot or cold, spicy, etc.) and practices, the study of which is known as gastronomy. Many cultures have diversified their foods by means of preparation, cooking methods, and manufacturing. This also includes a complex food trade which helps the cultures to economically survive by way of food, not just by consumption. Some popular types of ethnic foods include Italian, French, Japanese, Chinese, American, Cajun, Thai, and Indian cuisine.

Various cultures throughout the world study the dietary analysis of food habits. While evolutionarily speaking, as opposed to culturally, humans are omnivores, religion and social constructs such as morality, activism, or environmentalism will often affect which foods they will consume. Food is eaten and typically enjoyed through the sense of taste, the perception of flavour from eating and drinking. Certain tastes are more enjoyable than others, for evolutionary purposes.

Taste Perception

Animals, specifically humans, have five different types of tastes: sweet, sour, salty, bitter, and umami. As animals have evolved, the tastes that provide the most energy (sugar and fats) are the most pleasant to eat while others, such as bitter, are not enjoyable. Water, while important for survival, has no taste. Fats, on the other hand, especially saturated fats, are thicker and rich and are thus enjoyable to eat.

Sweet

Generally regarded as the most pleasant taste, sweetness is almost always caused by a type of simple sugar such as glucose or fructose, or disaccharides such as sucrose, a molecule combining glucose and fructose. Complex carbohydrates are long chains and thus do not have the sweet taste. Artificial sweeteners such as sucralose are used to mimic the sugar molecule, creating the sensation of sweet, without the calories. Other types of sugar include raw sugar, which is known for its amber color, as it is unprocessed. As sugar is vital for energy and survival, the taste of sugar is pleasant.

The stevia plant contains a compound known as steviol which, when extracted, has 300 times the sweetness of sugar while having minimal impact on blood sugar.

Sour

Sourness is caused by the taste of acids, such as vinegar or ethanol in alcoholic beverages. Sour foods include citrus, specifically lemons, limes, and to a lesser degree oranges. Sour is evolutionarily significant as it is a sign for a food that may have gone rancid due to bacteria. Many foods, however, are slightly acidic, and help stimulate the taste buds and enhance flavour.

Salty

Saltiness is the taste of alkali metal ions such as sodium and potassium. It is found in almost every food in low to moderate proportions to enhance flavour, although to eat pure salt is regarded as highly unpleasant. There are many different types of salt, with each having a different degree of saltiness, including sea salt, fleur de sel, kosher salt, mined salt, and grey salt. Other than enhancing flavour, its significance is that the body needs and maintains a delicate electrolyte balance, which is the kidney's function.

Salt may be iodized, meaning iodine has been added to it, a necessary nutrient that promotes thyroid function. Some canned foods, notably soups or packaged broths, tend to be high in salt as a means of preserving the food longer. Historically speaking, salt has been used as a meat preservative as salt promotes water excretion, thus working as a preservative. Similarly, dried foods also promote food safety.

Bitter

Bitterness is a highly unpleasant sensation characterized by having a sharp, pungent taste. Dark, unsweetened chocolate, caffeine, lemon rind, and some types of fruit are known to be bitter.

Umami

Umami, the Japanese word for delicious, is the least known in Western popular culture, but has a long tradition in Asian cuisine. Umami is the taste of glutamates, especially monosodium glutamate or MSG. It is characterized as savory, meaty, and rich in flavour. Salmon and mushrooms are foods high in umami. Meat and other animal by-products are described as having this taste.

Presentation

Food Presentation is the art of modifying, processing, arranging, or decorating food to enhance its aesthetic appeal.

The visual presentation of foods is often considered by chefs at many different stages of food preparation, from the manner of tying or sewing meats, to the type of cut used in chopping and slicing meats or vegetables, to the style of mold used in a poured dish. The food itself may be decorated as in elaborately iced cakes, topped with ornamental sometimes sculptural consumables, drizzled with sauces, sprinkled with seeds, powders, or other toppings, or it may be accompanied by edible or inedible garnishes.

The arrangement and overall styling of food upon bringing it to the plate is termed plating. Some common styles of plating include a 'classic' arrangement of the main item in the front of the plate with vegetables or starches in the back, a 'stacked' arrangement of the various items, or the main item leaning or 'shingled' upon a vegetable bed or side item. Item location on the plate is often referenced as for the face of a clock, with six o'clock the position closest to the diner.

It is known that when presented with food, the consumer "eats" first with their eyes, a universal psychological phenomenon. Food presented in a clean and appetizing way will encourage a good flavour, even if unsatisfactory.

Contrast in Texture

Texture plays a crucial role in the enjoyment of eating foods. Contrasts in textures, such as something crunchy in an otherwise smooth dish, may increase the appeal of eating it. Common examples include adding granola to yogurt, adding croutons to a salad or soup, and toasting bread to enhance its crunchiness for a smooth topping, such as jam or butter.

Contrast in Taste

Another universal phenomenon regarding food is the appeal of contrast in taste and presentation. Opposite flavours, such as sweet and saltiness, tend to go well together, such as in kettle corn and with nuts.

Food Preparation

While many foods can be eaten raw, many foods undergo some form of preparation for reasons of safety, palatability, texture, or flavour. At the simplest level this may involve washing, cutting, trimming, or adding other foods or ingredients, such as spices. It may also involve mixing, heating or cooling, pressure cooking, fermentation, or combination with other food. In a home, most food preparation takes place in a kitchen. Some preparation is done to enhance the taste or aesthetic appeal; other preparation may help to preserve the food; others may be involved in cultural identity. A meal is made up of food which is prepared to be eaten at a specific time and place.

Animal Preparation

The preparation of animal-based food usually involves slaughter, evisceration, hanging, portioning, and rendering. In developed countries, this is usually done outside the home in slaughterhouses, which are used to process animals en masse for meat production. Many countries regulate their slaughterhouses by law. For example, the United States has established the Humane Slaughter Act of 1958, which requires that an animal be stunned before killing. This act, like those in many countries, exempts slaughter in accordance to religious law, such as kosher, shechita, and dhabi%a halal. Strict interpretations of kashrut require the animal to be fully aware when its carotid artery is cut.

On the local level, a butcher may commonly break down larger animal meat into smaller manageable cuts, and pre-wrap them for commercial sale or wrap them to order in butcher paper. In addition, fish and seafood may be fabricated into smaller cuts by a fish monger. However fish butchery may be done on board a fishing vessel and quick-frozen for preservation of quality.

Cooking

Cooking is the process of preparing food with heat. Cooks select and combine ingredients using a wide range of tools and methods. In the process, the flavour, texture, appearance, and chemical properties of the ingredients can change. Cooking techniques and ingredients vary widely across the world, reflecting unique environmental, economic, and cultural traditions. Cooks themselves also vary widely in skill and training.

Preparing food with heat or fire is an activity unique to humans, and some scientists believe the advent of cooking played an important role in human evolution. Most anthropologists believe that cooking fires first developed around 250,000 years ago. The development of agriculture, commerce and transportation between civilizations in different regions offered cooks many new ingredients. New inventions and technologies, such as pottery for holding and boiling water, expanded cooking techniques. Some modern cooks apply advanced scientific techniques to food preparation.

History of Cooking

There is no clear evidence as to when cooking was invented. Primatologist Richard Wrangham stated that cooking was invented as far back as 1.8 million to 2.3 million years ago. Other researchers believe that cooking was invented as late as 40,000 or 10,000 years ago. Evidence of fire is inconclusive as wildfires started by lightning-strikes are still common in East Africa and other wild areas, and it is difficult to determine when fire was first used for cooking, as opposed to just being used for warmth or for keeping predators away. Most anthropologists contend that cooking fires began in earnest barely 250,000 years ago, when ancient hearths, earth ovens, burnt animal bones, and flint appear across Europe and the middle East. The only evidence of human use of fire more than two million years ago is burnt earth with human remains, which most anthropologists consider coincidence rather than evidence of intentional.

However, some Fire-cracked rock, such as that in Central Texas (United States) are burned rock middens, or enormous piles fire-damaged rock dated to c. 3,500 years ago. These may represent the remains of earth ovens used in cooking since they contain evidence of *Dasylirion wheeleri* bulbs and other plants. In Great Britain similar Neolithic, Bronze Age and Iron Age features exist, but are commonly called 'burnt mounds'.

Ingredients in Cooking

Most ingredients in cooking are derived from living things. Vegetables, fruits, grains and nuts come from plants, while meat, eggs, and dairy products come from animals. Mushrooms and the yeast used in baking are kinds of fungi. Cooks also utilize water and minerals such as salt. Cooks can also use wine, an alcohol-based liquid from the fermentation of juices of grapes or other fruits.

Naturally occurring ingredients contain various amounts of molecules called *proteins, carbohydrates* and *fats*. They also contain water and minerals. Cooking involves a manipulation of the chemical properties of these molecules.

Proteins

Edible animal material, including muscle, offal, milk, eggs and egg whites, contains substantial amounts of protein. Almost all vegetable matter (in particular legumes and seeds) also includes proteins, although generally in smaller amounts. These may also be a source of essential amino acids. When proteins are heated they become denatured and change texture. In many cases, this causes the structure of the material to become softer or more friable - meat becomes *cooked*. In some cases, proteins can form more rigid structures, such as the coagulation of albumen in egg whites. The formation of a relatively rigid but flexible matrix from egg white provides an important component of much cake cookery, and also underpins many desserts based on meringue.

Carbohydrates

Carbohydrates include the common sugar, sucrose (table sugar), a disaccharide, and such simple sugars as glucose (from the digestion of table sugar) and fructose (from fruit), and starches from sources such as cereal flour, rice, arrowroot, potato. The interaction of heat and carbohydrate is complex.

Long-chain sugars such as starch tend to break down into simpler sugars when cooked, while simple sugars can form syrups. If sugars are heated so that all water of crystallisation is driven off, then caramelization starts, with the sugar undergoing thermal decomposition with the formation of carbon, and other breakdown products producing caramel. Similarly, the heating of sugars and proteins elicits the Maillard reaction, a basic flavour-enhancing technique.

An emulsion of starch with fat or water can, when gently heated, provide thickening to the dish being cooked. In European cooking, a mixture of butter and flour called a roux is used to thicken liquids to make stews or sauces. In Asian cooking, a similar effect is obtained from a mixture of rice or corn starch and water. These techniques rely on the properties of starches to create simpler mucilaginous saccharides during cooking, which causes the familiar thickening of sauces. This thickening will break down, however, under additional heat.

Fats

Types of fat include vegetable oils and animal products such as butter and lard. Fats can reach temperatures higher than the boiling point of water, and are often used to conduct high heat to other ingredients, such as in frying or sautéing.

Water

Cooking often involves water which is frequently present as other liquids, both added in order to immerse the substances being cooked (typically water, stock or wine), and released from the foods themselves. Liquids are so important to cooking that the name of the cooking method used may be based on how the liquid is combined with the food, as in steaming, simmering, boiling, braising and blanching. Heating liquid in an open container results in rapidly increased evaporation, which concentrates the remaining flavour and ingredients - this is a critical component of both stewing and sauce making.

Vitamins and Minerals

Vitamins are materials required for normal metabolism but which the body cannot manufacture itself and which must therefore come from soil. Vitamins come from a number of sources including fresh fruit and vegetables (Vitamin C), carrots, liver (Vitamin A), cereal bran, bread, liver e (B vitamins), fish liver oil (Vitamin D) and fresh green vegetables (Vitamin K). Many minerals are also essential in small quantities including iron, calcium, magnesium and sulphur; and in very small quantities copper, zinc and selenium. The micronutrients, minerals, and vitamins in fruit and vegetables may be destroyed or eluted by cooking. Vitamin C is especially prone to oxidation during cooking and may be completely destroyed by protracted cooking.

Cooking Methods

Methods of Cooking

There are very many methods of cooking, most of which have been known since antiquity. These include baking, roasting, frying, grilling, barbecuing, smoking, boiling, steaming and braising. A more recent innovation is microwaving. Various methods use differing levels of heat and moisture and vary in cooking time. The method chosen greatly affects the end result. Some foods are more appropriate to some methods than others. Some major hot cooking techniques include:

Roasting

Roasting is a cooking method that uses dry heat, whether an open flame, oven, or other heat source. Roasting usually causes caramelization or Maillard browning of the surface of the food, which is considered a flavour enhancement. Roasting uses more indirect, diffused heat (as in an oven), and is suitable for slower cooking of meat in a larger, whole piece. Meats and most root and bulb vegetables can be roasted. Any piece of meat, especially red meat, that has been cooked in this fashion is called a roast. In addition, large uncooked cuts of meat are referred to as roasts. Roasting is a much slower method of cooking. A roast joint of meat can take one, two, even three hours to cook - the resulting meat is tender. Also, meats and vegetables prepared in this way are described as "roasted", e.g., roasted chicken or roasted squash.

Methods

For roasting, the food may be placed on a rack, in a roasting pan or, to ensure even application of heat, may be rotated on a spit or rotisserie. During oven roasting, hot air circulates around the meat, cooking all sides evenly. There are several theories for roasting meats correctly: low-temperature cooking, high-temperature cooking, and a combination of both. Each method can be suitable under appropriate circumstances.

- A low-temperature oven, 95 °C to 160 °C (200 °F to 325 °F), is best when cooking with large cuts of meat, turkey and whole chickens. This is not technically roasting temperature, but it is called slow-roasting. The benefit of slow-roasting an item is less moisture loss and a more tender product. At true roasting temperatures, 200 °C (400 °F) or more, the water inside the muscle is lost at a high rate.
- Cooking at high temperatures is beneficial if the cut is small enough—as in filet mignon or strip loin—to be finished cooking before the juices escape.
- The combination method uses high heat just at either the beginning or the end of the cooking process, with most of the cooking at a low temperature. This method produces the golden-brown texture and crust, but maintains more of the moisture than simply cooking at a high temperature, although the product will not be as moist as low-temperature cooking the whole time. Searing and then turning down to low is also beneficial when a dark crust and caramelized flavour is desired for the finished product. Note that searing in no way "locks in" moisture: moisture loss is simply a function of heat and time.

In general, in either case, the meat is removed from heat before it has finished cooking and left to sit for a few minutes, while the inside cooks further from the residual heat content, a phenomenon known as carry over cooking. The objective in any case is to retain as much moisture as possible, while providing the texture and color. During roasting, meats and vegetables are frequently basted on the surface with butter, lard, or oil to reduce the loss of moisture by evaporation. In recent times, plastic oven bags have become popular for roasts. These cut cooking times and reduce the loss of moisture during roasting, but reduce flavour development from Maillard browning. They are particularly popular for turkeys.

Until the late 19th century, roasting by dry heat in an oven was called *baking*. Roasting originally meant turning meat or a bird on a spit in front of a fire. It is one of the oldest forms of cooking known.

Traditionally recognized roasting methods consist only of baking and cooking over or near an open fire. Grilling is normally not technically a roast, since a grill (gridiron) is used (in English-speaking countries). Smoking differs from roasting because of the lower temperature and controlled smoke application.

Meat

Most meat roasts are large cuts of meat. Many roasts are tied with string prior to roasting, often using the reef knot or, in the more traditional sense, the packer's knot. Tying holds them together during roasting, keeping any stuffing inside, and keeps the roast in a round profile, which promotes even cooking. Prior to roasting in an oven, meat is generally "browned" by brief exposure to high temperature. This imparts a traditional flavour and color to the roast. Red meats such as beef, lamb, and venison, and certain game birds are often roasted to be "pink" or "rare", meaning that the center of the roast is still red. Due to food safety concerns, this practice is not recommended with pork and poultry. Although there is a growing fashion in some restaurants to serve "rose pork", temperature monitoring of the center of the roast is the only sure way to avoid foodborne disease.

Other

Roasting is a preferred method of cooking for most poultry, and certain cuts of beef, pork, or lamb. Some vegetables, such as potatoes, zucchini, pumpkin, turnips, parsnips, cauliflower, asparagus, squash, and peppers lend themselves to roasting as well. Roasted chestnuts are also a popular snack in winter.

Barbecuing – Grilling – Rotisserie – Searing

Baking

Baking is the technique of prolonged cooking of food by dry heat acting by convection, and not by radiation, normally in an oven, but also in hot ashes, or on hot stones. It is primarily used for the preparation of bread, cakes, pastries and pies, tarts, quiches, cookies and crackers. Such items are sometimes referred to as "baked goods," and are sold at a bakery. A person who prepares baked goods as a profession is called a baker. It is also used for the preparation of baked potatoes, baked apples, baked beans, some casseroles and pasta dishes such as lasagna, and various other foods, such as the pretzel.

Many commercial ovens are provided with two heating elements: one for baking, using convection and conduction to heat the food, and one for broiling or grilling, heating mainly by radiation. Meat may also be baked, but this is usually reserved for meatloaf, smaller cuts of whole meats, and whole meats that contain stuffing or coating such as breadcrumbs or buttermilk batter; larger cuts prepared without stuffing or coating are more often roasted, a similar process, using higher temperatures and shorter cooking times. Baking can sometimes be combined with grilling to produce a hybrid barbecue variant, by using both methods simultaneously or one before the other, cooking twice. Baking is connected to barbecuing because the concept of the masonry oven is similar to that of a smoke pit.

The baking process does not require any fat to be used to cook in an oven. Some makers of snacks such as potato chips or crisps have produced baked versions of their snack items as an alternative to the usual cooking method of deep-frying in an attempt to reduce the calorie or fat content of their snack products.

Baking Blind - Broiling - Flashbaking

Boiling

Boiling is the rapid vaporization of a liquid, which occurs when a liquid is heated to its boiling point, the temperature at which the vapor pressure of the liquid is equal to the pressure exerted on the liquid by the surrounding environmental pressure. While below the boiling point a liquid evaporates from its surface, at the boiling point vapor bubbles come from the bulk of the liquid. For this to be possible, the vapor pressure must be sufficiently high to win the atmospheric pressure, so that the bubbles can be "inflated". Thus, the difference between evaporation and boiling is "mechanical", rather than thermodynamical. The boiling point is lowered when the pressure of the surrounding atmosphere is reduced, for example by the use of a vacuum pump or at high altitudes. Boiling occurs in three characteristic stages, which are *nucleate, transition* and *film boiling*. These stages generally take place from low to high heating surface temperatures, respectively.

Boiling Stages

Nucleate Boiling

Nucleate boiling is characterized by the growth of bubbles on a heated surface, which rise from discrete points on a surface, whose temperature is only slightly above the liquid's. In general, the number of nucleation sites are increased by an increasing surface temperature.

An irregular surface of the boiling vessel (i.e. increased surface roughness) can create additional nucleation sites, while an exceptionally smooth surface, such as glass, lends itself to superheating. Under these conditions, a heated liquid may show boiling delay and the temperature may go somewhat above the boiling point without boiling.

Transition Boiling

Transition boiling may be defined as the unstable boiling, which occurs at surface temperatures between the maximum attainable in nucleate and the minimum attainable in film boiling.

The formation of bubbles in a heated liquid is a complex physical process which often involves cavitation and acoustic effects, such as the broad-spectrum hiss one hears in a kettle not yet heated to the point where bubbles boil to the surface.

Film Boiling

If a surface heating the liquid is significantly hotter than the liquid then film boiling will occur, where a thin layer of vapor, which has low thermal conductivity insulates the surface. This condition of a vapor film insulating the surface from the liquid characterizes *film boiling*.

Applications

Distillation

In distillation, boiling is used to separate mixtures. This is possible because the vapor rising from a boiling fluid generally has a ratio of components different from that in the liquid. It must be boiled for a full 10 minutes in order to kill all pathogens and be considered sterile.

Boiling for Water Sterilization

Boiling can be used as a method of water disinfection but is only advocated as an emergency water treatment method, or as a method of portable water purification in rural or wilderness settings without access to a potable water infrastructure. Although bringing water to the boil is effective in killing or inactivating most bacteria, viruses and pathogens, some are resistant and can survive several minutes boiling especially at high altitude where the temperature at which water boils is reduced.

Boiling in Cooking

In cooking, *boiling* is the method of cooking food in boiling water, or other water-based liquid such as stock or milk. Simmering is gentle boiling, while in poaching the cooking liquid moves but scarcely bubbles. Boiling is a very harsh technique of cooking. Delicate foods such as fish cannot be cooked in this fashion because the bubbles can damage the food. Foods such as red meat, chicken, and root vegetables can be cooked with this technique because of their tough texture.

The open-air boiling point of water is typically considered to be 100 °C or 212 °F. Pressure and a change in composition of the liquid may alter the boiling point of the liquid. For this reason, high elevation cooking generally takes longer since boiling point is a function of atmospheric pressure. In Denver, Colorado, which is at an elevation of about one mile, water boils at approximately 95 °C. Depending on the type of food and the elevation, the boiling water may not be hot enough to cook the food properly. Similarly, increasing the pressure as in a pressure cooker raises the temperature of the contents above the open air boiling point.

Adding a water soluble substance, such as salt or sugar also increases the boiling point. This is called boiling-point elevation. However, the effect is very small, and the boiling point will be increased by an insignificant amount. Due to variations in composition and pressure, the boiling point of water is almost never exactly 100 °C, but rather close enough for cooking.

Bringing water to a boil is generally done by applying maximal heat, then shutting off when the water has come to a boil, which is known as bang–bang control. Keeping water at or below a boil requires more careful control of temperature, particularly by using feedback.

In places where the available water supply is contaminated with disease-causing bacteria, boiling water and allowing it to cool before drinking it is practiced as a valuable health measure. Boiling water for a few minutes kills most bacteria, amoeba, and other microbial pathogens. It thus can help prevent cholera, dysentery, and other diseases caused by microorganisms. Foods suitable for boiling include vegetables, starchy foods such as rice, noodles and potatoes, eggs, meats, sauces, stocks and soups. Boiling has several advantages. It is safe and simple, and it is appropriate for large-scale cookery Older, tougher, cheaper cuts of meat and poultry can be made digestible. Nutritious, well flavoured stock is produced. Also, maximum color and nutritive value is retained when cooking green vegetables, provided boiling time is kept to the minimum.

On the other hand, there are several disadvantages. There is a loss of soluble vitamins from foods to the water (if the water is discarded), and some boiled foods can look unattractive. Boiling can also be a slow method of cooking food.

Boiling can be done in several ways: The food can be placed into already rapidly boiling water and left to cook, the heat can be turned down and the food can be simmered; or the food can also be placed into the pot, and cold water may be added to the pot. This may then be boiled until the food is satisfactory. Water on the outside of a pot, i.e. a wet pot, increases the time it takes the pot of water to boil. The pot will heat at a normal rate once all excess water on the outside of the pot evaporates.

Levels of Boiling

In Chinese cuisine, particularly tea brewing, one distinguishes five stages of boiling: *"shrimp eyes,* the first tiny bubbles that start to appear on the surface of the kettle water, *crab eyes,* the secondary, larger bubbles, then *fish eyes,* followed by *rope of pearls,* and finally *raging torrent* [rolling boil]".

In detail:

shrimp eyes: about 70-80 °C (155–175 °F) – separate bubbles, rising to top.

crab eyes: about 80 °C (175 °F) – streams of bubbles.

fish eyes: about 80-90 °C (175–195 °F) – larger bubbles.

rope of pearls: about 90-95 °C (195–205 °F) – steady streams of large bubbles.

raging torrent: rolling boil, swirling and roiling.

Blanching - Braising - Coddling - Double steaming - Infusion - Poaching - Pressure cooking - Simmering - Steaming - Steeping - Stewing - Vacuum flask cooking.

Frying

Frying is the cooking of food in oil or fat, a technique that originated in ancient Egypt around 2500 BC. Chemically, oils and fats are the same, differing only in melting point, but the distinction is only made when needed. In commerce, many fats are called oils by custom, e.g. palm oil and coconut oil, which are solid at room temperature.

History

Frying is believed to have originated in ancient Egypt around 2500 BC.

Details

Fats can reach much higher temperatures than water at normal atmospheric pressure. Through frying, one can sear or even carbonize the surface of foods while caramelizing sugars. The food is cooked much more quickly and has a characteristic crispness and texture. Depending on the food, the fat will penetrate it to varying degrees, contributing richness, lubricity, and its own flavour.

Frying techniques vary in the amount of fat required, the cooking time, the type of cooking vessel required, and the manipulation of the food. Sautéing, stir frying, pan frying, shallow frying, and deep frying are all standard frying techniques.

Sautéing and stir-frying involve cooking foods in a thin layer of fat on a hot surface, such as a frying pan, griddle, wok, or sauteuse. Stir frying involves frying quickly at very high temperatures, requiring that the food be stirred continuously to prevent it from adhering to the cooking surface and burning.

Shallow frying is a type of pan frying using only enough fat to immerse approximately one-third to one-half of each piece of food; fat used in this technique is typically only used once.

Deep-frying, on the other hand, involves totally immersing the food in hot oil, which is normally topped up and used several times before being disposed. Deep-frying is typically a much more involved process, and may require specialized oils for optimal results.

Deep frying is now the basis of a very large and expanding worldwide industry. Fried products have consumer appeal in all age groups, and the process is quick, can easily be made continuous for mass production, and the food emerges sterile and dry, with a relatively long shelf life. The end products can then be easily packaged for storage and distribution. Examples are potato chips, french fries, nuts, doughnuts, instant noodles, etc.

Smoking

Smoking is the process of flavouring, cooking, or preserving food by exposing it to the smoke from burning or smoldering plant materials, most often wood. Meats and fish are the most common smoked foods, though cheeses, vegetables, and ingredients used to make beverages such as whisky, Rauchbier and *lapsang souchong* tea are also smoked.

In Europe, alder is the traditional smoking wood, but oak is more often used now, and beech to a lesser extent. In North America, hickory, mesquite, oak, pecan, alder, maple, and fruit-tree woods, such as apple, cherry and plum, are commonly used for smoking. Other fuels besides wood can also be employed, sometimes with the addition of flavouring ingredients. Chinese tea-smoking uses a mixture of uncooked rice, sugar, and tea, heated at the base of a wok. Some North American ham and bacon makers smoke their products over burning corncobs. Peat is burned to dry and smoke the barley malt used to make whisky and some beers. In New Zealand, sawdust from the native manuka (tea tree) is commonly used for hot smoking fish. In Iceland, dried sheep dung is used to cold smoke fish, lamb, mutton and whale, resulting in a unique and rather strongly smoked flavour.

Historically, farms in the western world included a small building termed the *smokehouse*, where meats could be smoked and stored. This was generally well-separated from other buildings both because of the fire danger and because of the smoke emanations.

Types

- Cold smoking can be used as a flavour enhancer for items such as chicken breasts, beef, pork chops, salmon, scallops, and steak. The item can be cold smoked for just long enough to give some flavour. Some cold smoked foods are baked, grilled, roasted, or sautéed before eating. Smokehouse temperatures for cold smoking are below 100 °F (38 °C). In this temperature range, foods take on a smoked flavour, but remain relatively moist. Cold smoking does not cook foods.
- Hot smoking exposes the foods to smoke and heat in a controlled environment. Although foods that have been hot smoked are often reheated or cooked, they are typically safe to eat without further cooking. Hams and ham hocks are fully cooked once they are properly smoked. Hot smoking occurs within the range of 165 °F (74 °C) to 185 °F (85 °C). Within this temperature range, foods are fully cooked, moist, and flavourful. If the smoker is allowed to get hotter than 185 °F (85 °C), the foods will shrink excessively, buckle, or even split. Smoking at high temperatures also reduces yield, as both moisture and fat are "cooked" away.
- Smoke roasting or smoke baking refers to any process that has the attributes of smoking combined with either roasting or baking. This smoking method is sometimes referred to as "barbecuing", "pit baking", or "pit roasting". It may be done in a smoke roaster, closed wood-fired masonry oven or barbecue pit, any smoker that can reach above 250 °F (121 °C), or in a conventional oven by placing a pan filled with hardwood chips on the floor of the oven so the chips smolder and produce a

smokebath. However, this should only be done in a well-ventilated area to prevent carbon monoxide poisoning.

Wood Smoke

Hardwoods are made up mostly of three materials: cellulose, hemicellulose, and lignin. Cellulose and hemicellulose are the basic structural material of the wood cells; lignin acts as a kind of cell-bonding glue. Some softwoods, especially pines and firs, hold significant quantities of resin, which produces a harsh-tasting soot when burned; these woods are not often used for smoking.

Cellulose and hemicellulose are aggregate sugar molecules; when burnt, they effectively caramelize, producing carbonyls, which provide most of the color components and sweet, flowery, and fruity aromas.

Lignin, a highly complex arrangement of interlocked phenolic molecules, also produces a number of distinctive aromatic elements when burnt, including smoky, spicy, and pungent compounds such as guaiacol, phenol, and syringol, and sweeter scents such as the vanilla-scented vanillin and clove-like isoeugenol. Guaiacol is the phenolic compound most responsible for the "smokey" taste, while syringol is the primary contributor to smokey aroma. Wood also contains small quantities of proteins, which contribute roasted flavours. Many of the odor compounds in wood smoke, especially the phenolic compounds, are unstable, dissipating after a few weeks or months.

A number of wood smoke compounds act as preservatives. Phenol and other phenolic compounds in wood smoke are both antioxidants, which slow rancidification of animal fats, and antimicrobials, which slow bacterial growth. Other antimicrobials in wood smoke include formaldehyde, acetic acid, and other organic acids, which give wood smoke a low pH—about 2.5. Some of these compounds are toxic to people as well, and may have health effects in the quantities found in cooking applications.

Since different species of trees have different ratios of components, various types of wood do impart a different flavour to food. Another important factor is the temperature at which the wood burns. High-temperature fires see the flavour molecules broken down further into unpleasant or flavorless compounds.

The optimal conditions for smoke flavour are low, smoldering temperatures between 570 and 750 °F (299 and 399 °C). This is the temperature of the burning wood itself, not of the smoking environment, which uses much lower temperatures. Woods that are high in lignin content tend to burn hot; to keep them smoldering requires restricted oxygen supplies or a high moisture content. When smoking using wood chips or chunks, the combustion temperature is often raised by soaking the pieces in water before placing them on a fire.

Types of Smoker

There are a few basic types of smoker designs, each with their own advantages and disadvantages.

Charcoal Smokers

Offset Smokers

The main characteristics of the offset smoker are that the cooking chamber is usually cylindrical in shape, with a shorter, smaller diameter cylinder attached to the bottom of one end for a firebox. To cook the meat, a small fire is lit in the firebox, where airflow is tightly controlled.

The heat and smoke from the fire is drawn through a connecting pipe or opening into the cooking chamber. The heat and smoke cook and flavour the meat before escaping through an exhaust vent at the opposite end of the cooking chamber. Most manufacturers' models are based on this simple but effective design, and this is what most people picture when they think of a "BBQ smoker." Even large capacity commercial units use this same basic design of a separate, smaller fire box and a larger cooking chamber.

The UDS

The Upright Drum Smoker (also referred to as an Ugly Drum Smoker or UDS) is exactly what its name suggests; an upright steel drum that has been modified for the purpose of pseudo-indirect hot smoking. There are many ways to accomplish this, but the basics include the use of a complete steel drum, a basket to hold charcoal near the bottom, and cooking rack (or racks) near the top; all covered by a vented lid of some sort. They have been built using many different sizes of steel drums (30 gallon, 55 gallon, and 85 gallon for example), but the most popular size is the common 55 gallon drum.

This design is similar to smoking with indirect heat due to the distance from the coals and the racks (typically 24"). The temperatures used for smoking are controlled by limiting the amount of air intake at the bottom of the drum, and allowing a similar amount of exhaust out of vents in the lid. UDSs are very efficient with fuel consumption and flexible in their abilities to produce proper smoking conditions, with or without the use of a water pan or drip pan. Most UDS builders/users would say a water pan defeats the true pit BBQ nature of the UDS, as the drippings from the smoked meat should land on the coals, burning up, and imparting a unique flavour one cannot get with a water pan.

Vertical Water Smoker

A vertical water smoker (also referred to as a bullet smoker because of its shape) is a variation of the upright drum smoker. It uses charcoal or wood to generate smoke and heat, and contains a water bowl between the fire and the cooking grates. The water bowl serves to hold the temperature down and also to add humidity to the smoke chamber.

In addition, the bowl catches any drippings from the meat that may cause a flare up. Vertical water smokers are extremely temperature stable and require very little adjustment once the desired temperature has been reached. Because of their relatively low cost and stable temperature, they are sometimes used in barbecue competitions where propane and electric smokers are not allowed.

Propane Smoker

A propane smoker is designed to allow the smoking of meat in a somewhat more controlled environment. The primary differences are the sources of heat and of the smoke. In a propane smoker, the heat is generated by a gas burner directly under a steel or iron box containing the wood or charcoal that provides the smoke. The steel box has few vent holes, on the top of the box only. By starving the heated wood of oxygen, it smokes instead of burning. Any combination of woods and charcoal may used. This method uses less wood.

Smoke Box Method

This more traditional method uses a two box system: The fire box and the food box. The fire box is typically adjacent or under the cooking box, and can be controlled to a finer degree. The heat and smoke from the fire box exhausts into the food box, where it is used to cook and smoke the meat. These may be as simple as an electric heating element with a pan of wood chips placed on it, although more advanced models have finer temperature controls.

Commercial Smoke House

Commercial smokehouses, mostly made from stainless steel, have independent systems for smoke generation and cooking. Smoke generators use friction, an electric coil or a small flame to ignite sawdust on demand. Heat from steam coils or gas flames is balanced with live steam or water sprays to control the temperature and humidity. Elaborate air handling systems reduce hot or cold spots, to reduce variation in the finished product. Racks on wheels or rails are used to hold the product and facilitate movement.

Preservation

Smoke is an antimicrobial and antioxidant, but smoke alone is insufficient for preserving food in practice, unless combined with another preservation method. The main problem is the smoke compounds adhere only to the outer surfaces of the food; smoke does not actually penetrate far into meat or fish. In modern times, almost all smoking is carried out for its flavour. Artificial smoke flavouring can be purchased as a liquid to mimic the flavour of smoking, but not its preservative qualities.

In the past, smoking was a useful preservation tool, in combination with other techniques, most commonly salt-curing or drying. In some cases, particularly in climates without much hot sunshine, smoking was simply an unavoidable side effect of drying over a fire.

For some long-smoked foods, the smoking time also served to dry the food. Drying, curing, or other techniques can render the interior of foods inhospitable to bacterial life, while the smoking gives the vulnerable exterior surfaces an extra layer of protection.

For oily fish smoking is especially useful, as its antioxidant properties delay surface fat rancidification. (Interior fat is not as exposed to oxygen, which is what causes rancidity.) Some heavily-salted, long-smoked fish can keep without refrigeration for weeks or months. Such heavily-preserved foods usually require a treatment such as boiling in fresh water to make them palatable before eating.

Cancer Risk

"Of various sources of N-nitroso compounds, intake of smoked and salted fish was significantly (RR = 2.58, 95% CI 1.21 " 5.51) and intake of cured meat was non-significantly (RR = 1.84, 95% CI 0.98–3.47) associated with risk of colorectal cancer."

Some Smoked Foods

Many foods can be smoked. Some of the more common are listed here.

Beverages

- *Lapsang souchong* tea leaves are smoked and dried over pine or cedar fires
- Malt beverages

- o The malt used to make whisky
- o Rauchbier (smoked beer).

Fruit
- Capsicums: chipotles (smoked, ripe jalapeño peppers), paprika
- Prunes (dried plums) can be smoked while drying
- *Wumei* are smoked ume fruit.

Protein
- Cheeses
- Fish :
  - o Eel popular in eastern/northern Europe and Japan, where it is called *unagi*
  - o Grimsby traditional smoked fish (cod and haddock)
  - o Haddock and Arbroath Smokies (haddock)
  - o Kippers and bloater (herring)
  - o Salmon.
- Beef :
  - o Traditionally-prepared jerky
  - o Pastrami (pickled, spiced and smoked beef brisket).
- Pork :
  - o Bacon
  - o Prosciutto
  - o Ham
  - o Various sausages.

Spices
- Paprika
- Salt.

Food Safety

Food safety is a scientific discipline describing handling, preparation, and storage of food in ways that prevent foodborne illness. This includes a number of routines that should be followed to avoid potentially severe health hazards. Food can transmit disease from person to person as well as serve as a growth medium for bacteria that can cause food poisoning. Debates on genetic food safety include such issues as impact of genetically modified food on health of further generations and genetic pollution of environment, which can destroy natural biological diversity. In developed countries there are intricate standards for food preparation, whereas in lesser developed countries the main issue is simply the availability of adequate safe water, which is usually a critical item. In theory food poisoning is 100% preventable.

Key Principles

Five Key Principles

The five key principles of food hygiene, according to WHO, are:
1. Prevent contaminating food with pathogens spreading from people, pets, and pests.
2. Separate raw and cooked foods to prevent contaminating the cooked foods.
3. Cook foods for the appropriate length of time and at the appropriate temperature to kill pathogens.
4. Store food at the proper temperature.
5. Use safe water and raw materials.

ISO 22000

ISO 22000 is a standard developed by the International Organization for Standardization dealing with food safety. This is a general derivative of ISO 9000. ISO 22000 standard: The ISO 22000 international standard specifies the requirements for a food safety management system that involves interactive communication, system management, prerequisite programs, HACCP principles.

Incidence

A 2003 World Health Organization (WHO) report concluded that about 40% of reported food poisoning outbreaks in the WHO European Region occur in private homes. According to the WHO and CDC, in the USA alone, annually, there are 76 million cases of foodborne illness leading to 325,000 hospitalizations and 5,000 deaths.

Regulatory Agencies

Australia

Australian Food Authority is working toward ensuring that all food businesses implement food safety systems to ensure food is safe to consume in a bid to halt the increasing incidence of food poisoning, this includes basic food safety training for at least one person in each business. Smart business operators know that basic food safety training improves the bottom line, staff take more pride in their work; there is less waste; and customers can have more confidence in the food they consume. Food Safety training in units of competence from a relevant training package, must be delivered by a Registered Training Organization (RTO) to enable staff to be issued with a nationally-recognised unit of competency code on their certificate. Generally this training can be completed in less than one day. Training options are available to suit the needs of everyone. Training may be carried out in-house for a group, in a public class, via correspondence or on-line. Basic food safety training includes:

- Understanding the hazards associated with the main types of food and the conditions to prevent the growth of bacteria which can cause food poisoning
- The problems associated with product packaging such as leaks in vacuum packs, damage to packaging or pest infestation, as well as problems and diseases spread by pests.
- Safe food handling. This includes safe procedures for each process such as receiving, re-packing, food storage, preparation and cooking, cooling and re-heating, displaying products, handling products when serving customers, packaging, cleaning and sanitizing, pest control, transport and delivery. Also the causes of cross contamination.
- Catering for customers who are particularly at risk of food-borne illness, including allergies and intolerance.
- Correct cleaning and sanitizing procedures, cleaning products and their correct use, and the storage of cleaning items such as brushes, mops and cloths.
- Personal hygiene, hand washing, illness, and protective clothing.

People responsible for serving unsafe food can be liable for heavy fines under this new leglislation, consumers are pleased that industry will be forced to take food safety seriously.

China

Food safety is a growing concern in Chinese agriculture. The Chinese government oversees agricultural production as well as the manufacture of food packaging, containers, chemical additives, drug production, and business regulation. In recent years, the Chinese government attempted to consolidate food regulation with the creation of the State Food and Drug Administration in 2003, and officials have also been under increasing public and international pressure to solve food safety problems. However, it appears that regulations are not well known by the trade. Labels used for "green" food, "organic" food and "pollution-free" food are not well recognized by traders and many are unclear about their meaning. A survey by the World Bank found that supermarket managers had difficulty in obtaining produce that met safety requirements and found that a high percentage of produce did not comply with established standards.

Traditional marketing systems, whether in China or the rest of Asia, presently provide little motivation or incentive for individual farmers to make improvements to either quality or safety as their produce tends to get grouped together with standard products as it progresses through the marketing channel. Direct linkages between farmer groups and traders or ultimate buyers, such as supermarkets, can help avoid this problem. Governments need to improve the condition of many markets through upgrading management and reinvesting market fees in physical infrastructure. Wholesale markets need to investigate the feasibility of developing separate sections to handle fruits and vegetables that meet defined safety and quality standards.

European Union

The parliament of the European Union (EU) makes legislation in the form of directives and regulations, many of which are mandatory for member states and which therefore must be incorporated into individual countries' national legislation. As a very large organisation that exists to remove barriers to trade between member states, and into which individual member states have only a proportional influence, the outcome is often seen as an excessively bureaucratic 'one size fits all' approach. However, in relation to food safety the tendency to err on the side of maximum protection for the consumer may be seen as a positive benefit. The EU parliament is informed on food safety matters by the European Food Safety Authority.

Individual member states may also have other legislation and controls in respect of food safety, provided that they do not prevent trade with other states, and can differ considerably in their internal structures and approaches to the regulatory control of food safety.

Germany

The Federal Ministry of Food, Agriculture and Consumer Protection (BMELV) is a Federal Ministry of the Federal Republic of Germany. History: Founded as Federal Ministry of Food, Agriculture and Foresting in 1949, this name did not change until 2001. Then the name changed to Federal Ministry of Consumer Protection, Food and

Agriculture. At the 22nd of November 2005, the name got changed again to its current state: Federal Ministry of Food, Agriculture and Consumer Protection. The reason for this last change was that all the resorts should get equal ranking which was achieved by sorting the resorts alphabetically. Vision: A balanced and healthy diet with safe food, distinct consumer rights and consumer information for various areas of life, and a strong and sustainable agriculture as well as perspectives for our rural areas are important goals of the Federal Ministry of Food, Agriculture and Consumer Protection (BMELV). The Federal Office of Consumer Protection and Food Safety is under the control of the Federal Ministry of Food, Agriculture and Consumer Protection. It exercises several duties, with which it contributes to safer food and thereby intensifies health-based consumer protection in Germany. Food can be manufactured and sold within Germany without a special permission, as long as it does not cause any damage on consumers' health and meets the general standards set by the legislation. However, manufacturers, carriers, importers and retailers are responsible for the food they pass into circulation. They are obliged to ensure and document the safety and quality of their food with the use of in-house control mechanisms.

Hong Kong

In Hong Kong SAR, the Centre for Food Safety is in charge of ensuring food sold is safe and fit for consumption.

Pakistan

Pakistan does not have an integrated legal framework but has a set of laws, which deals with various aspects of food safety. These laws, despite the fact that they were enacted long time ago, have tremendous capacity to achieve at least minimum level of food safety. However, like many other laws, these laws remain very poorly enforced. There are four laws that specifically deal with food safety. Three of these laws directly focus issues related to food safety, while the fourth one namely Pakistan Standards and Quality Control Authority Act, is indirectly relevant to food safety. The Pure Food Ordinance, 1960 consolidates and amends the law in relation to the preparation and the sale of foods. All provinces and some northern areas have adopted this law with certain amendments. Its aim is to ensure purity of food being supplied to people in the market and, therefore, provides for preventing adulteration. The Pure Food Ordinance 1960 does not apply to cantonment areas. There is separate law for cantonments called "The Cantonment Pure Food Act, 1966". There is no substantial difference between the Pure Food Ordinance 1960 and The Cantonment Pure Food Act. Even the rules of operation are very much similar. Pakistan Hotels and Restaurant Act, 1976 applies to all hotels and restaurants in Pakistan and seeks to control and regulate the rates and standard of service(s) by hotels and restaurants.

In addition to other provisions, under section 22(2), the sale of food or beverages that are contaminated, not prepared hygienically or served in utensils that are not hygienic or clean is an offense. There are no express provisions for consumer complaints in the Pakistan Restaurants Act, 1976, Pakistan Penal Code, 1860 and Pakistan Standards and Quality Control Authority Act, 1996. However, the laws do not prevent citizens from lodging complaints with the concerned government officials. However, the consideration and handling of complaints is a matter of discretion of the officials.

South Korea

Korea Food & Drug Administration: Korea Food & Drug Administration (KFDA) is working for food safety since 1945. It is part of the Government of South Korea.

IOAS-Organic Certification Bodies Registered in KFDA: "Organic" or related claims can be labelled on food products when organic certificates are considered as valid by KFDA. KFDA admits organic certificates which can be issued by 1) IFOAM (International Federation of Organic Agriculture Movement) accredited certification bodies 2) Government accredited certification bodies - 328 bodies in 29 countries have been registered in KFDA.

Food Import Report: According to Food Import Report, it is supposed to report or register what you import. Competent authority is as followed:

* Imported Agricultural Products, Processed Foods, Food Additives, Utensils, Containers & Packages or Health Functional Foods ® KFDA (Korea Food and Drug Administration)
* Imported Livestock, Livestock products (including Dairy products) ® NVRQS (National Veterinary Research and Quarantine Service) -

    Packaged meat, milk & dairy products (butter, cheese), hamburger patties, meat ball and other processed products which are stipulated by Livestock Sanitation Management Act.

* Imported Marine products ® NFIS (National Fisheries Products Quality Inspection Service) - Fresh, chilled, frozen, salted, dehydrated, eviscerated marine produce which can be recognized its characteristics.

National Institute of Food and Drug Safety Evaluation

National Institute of Food and Drug Safety Evaluation (NIFDS) is functioning as well. The National Institute of Food and Drug Safety Evaluation is a national organization for toxicological tests and research. Under the Korea Food & Drug Administration, the Institute performs research on toxicology, pharmacology, and risk analysis of foods, drugs, and their additives. The Institute strives primarily to understand important biological triggering mechanisms and improve assessment methods of human exposure, sensitivities, and risk by (1) conducting basic, applied, and policy research that closely examines biologically triggering harmful effects on the regulated products such as foods, food additives, and drugs, and (2) operating the national toxicology program for the toxicological test development and inspection of hazardous chemical ubstances assessments. The Institute ensures safety by (1) investigation and research on safety by its own researchers, (2) contract research by external academicians and research centers.

United States

The US food system is regulated by numerous federal, state and local officials. Although the US food safety system is one of the best in the world, it is lacking in "organization, regulatory tools, and resources to address food borne illness."

Federal Level Regulation

The Food and Drug Administration publishes the Food Code, a model set of guidelines and procedures that assists food control jurisdictions by providing a scientifically sound technical and legal basis for regulating the retail and food service industries, including restaurants, grocery stores and institutional foodservice providers such as nursing homes. Regulatory agencies at all levels of government in the United States use the FDA Food Code to develop or update food safety rules in their jurisdictions that are consistent with national food regulatory policy. According to the FDA, 48 of 56 states and territories, representing 79% of the U.S. population, have adopted food codes patterned after one of the five versions of the Food Code, beginning with the 1993 edition. In the United States, federal regulations governing food safety are fragmented and complicated, according to a February 2007 report from the Government Accountability Office. There are 15 agencies sharing oversight responsibilities in the food safety system, although the two primary agencies are the U.S. Department of Agriculture (USDA) Food Safety and Inspection Service (FSIS), which is responsible for the safety of meat, poultry, and processed egg products, and the Food and Drug Administration (FDA), which is responsible for virtually all other foods.

The Food Safety and Inspection Service has approximately 7,800 inspection program personnel working in nearly 6,200 federally inspected meat, poultry and processed egg establishments.

FSIS is charged with administering and enforcing the Federal Meat Inspection Act, the Poultry Products Inspection Act, the Egg Products Inspection Act, portions of the Agricultural Marketing Act, the Humane Slaughter Act, and the regulations that implement these laws. FSIS inspection program personnel inspect every animal before slaughter, and each carcass after slaughter to ensure public health requirements are met. In fiscal year (FY) 2008, this included about 50 billion pounds of livestock carcasses, about 59 billion pounds of poultry carcasses, and about 4.3 billion pounds of processed egg products. At U.S. borders, they also inspected 3.3 billion pounds of imported meat and poultry products.

Industry Pressure

There have been concerns over the efficacy of safety practices and food industry pressure on U.S. regulators. A study reported by Reuters found that "the food industry is jeopardizing U.S. public health by withholding information from food safety investigators or pressuring regulators to withdraw or alter policy designed to protect consumers". A survey found that 25% of U.S. government inspectors and scientists surveyed have experienced during the past year corporate interests forcing their food safety agency to withdraw or to modify agency policy or action that protects consumers.

Scientists have observed that management undercuts field inspectors who stand up for food safety against industry pressure. According to Dr. Dean Wyatt, a USDA veterinarian who oversees federal slaughter house inspectors, "Upper level management does not adequately support field inspectors and the actions they take to protect the food supply.

Not only is there lack of support, but there's outright obstruction, retaliation and abuse of power."

Food Processing

State and Local Regulation

A number of U.S. states have their own meat inspection programs that substitute for USDA inspection for meats that are sold only in-state. Certain state programs have been criticized for undue leniency to bad practices. However, other state food safety programs supplement, rather than replace, Federal inspections, generally with the goal of increasing consumer confidence in the state's produce. For example, state health departments have a role in investigating outbreaks of food-borne disease bacteria, as in the case of the 2006 outbreak of *Escherichia coli* O157:H7 (bad *E. coli* bacteria) from processed spinach. Health departments also promote better food processing practices to eliminate these threats. In addition to the US Food and Drug Administration, several states that are major producers of fresh fruits and vegetables (including California, Arizona and Florida) have their own state programs to test produce for pesticide residues.

Restaurants and other retail food establishments fall under state law and are regulated by state or local health departments. Typically these regulations require official inspections of specific design features, best food-handling practices, and certification of food handlers. In some places a letter grade or numerical score must be prominently posted following each inspection. In some localities, inspection deficiencies and remedial action are posted on the Internet.

Manufacturing Control

HACCP Guidelines

The UK Food Standards Agency publishes recommendations as part of its Hazard Analysis and Critical Control Points (HACCP) programme. The relevant guidelines state that:

"Cooking food until the Core Temperature is 75 °C or above will ensure that harmful bacteria are destroyed.

However, lower cooking temperatures are acceptable provided that the Core Temperature is maintained for a specified period of time as follows :

- 60 °C for a minimum of 45 minutes
- 65 °C for a minimum of 10 minutes
- 70 °C for a minimum of 2 minutes".

Previous guidance from a leaflet produced by the UK Department Of Health "Handling Cooked Meats Safely A Ten Point Plan" also allowed for:

- "75 °C for a minimum of 30 seconds
- 80 °C for a minimum of 6 seconds".

Note that recommended cooking conditions are only appropriate if initial bacterial numbers in the uncooked food are small. Cooking does not overcome poor hygiene.

Consumer Labelling

United Kingdom

Food stuffs in the UK have one of two labels to indicate the nature of the deterioration of the product and any subsequent health issues. EHO Food Hygiene certification is required to prepare and distribute food. While there is no specified expiry date of such a qualification the changes in legislation it is suggested to update every five years. Best before indicates a future date beyond which the food product *may* lose quality in terms of taste or texture amongst others, but does not imply any serious health problems if food is consumed beyond this date (within reasonable limits). Use by indicates a legal date beyond which it is not permissible to sell a food product (usually one that deteriorates fairly rapidly after production) due to the potential serious nature of consumption of pathogens. Leeway is sometimes provided by producers in stating display until dates so that products are not at their limit of safe consumption on the actual date stated (this latter is voluntary and not subject to regulatory control). This allows for the variability in production, storage and display methods.

United States

With the exception of infant formula and baby foods which must be withdrawn by their expiration date, Federal law does not require expiration dates. For all other foods, except dairy products in some states, freshness dating is strictly voluntary on the part of manufacturers. In response to consumer demand, perishable foods are typically labelled with a Sell by date. It is up to the consumer to decide how long after the Sell by date a package is usable. Other common dating statements are Best if used by, Use-by date, Expiration date, Guaranteed fresh <date>, and Pack date.

Australia and New Zealand

Guide to Food Labelling and Other Information Requirements: This guide provides background information on the general labelling requirements in the Code. The information in this guide applies both to food for retail sale and to food for catering purposes. Foods for catering purposes means those foods for use in restaurants, canteens, schools, caterers or self-catering institutions, where food is offered for immediate consumption. Labelling and information requirements in the new Code apply both to food sold or prepared for sale in Australia and New Zealand and food imported into Australia and New Zealand. Warning and Advisory Declarations, Ingredient Labelling, Date Marking, Nutrition Information Requirements, Legibility Requirements for Food Labels, Percentage Labelling, Information Requirements for Foods Exempt from Bearing a Label.

Issues Associated with Sell by/Use by Dates

According to the UK's Waste & Resources Action Programme, 33% percent of all food produced is wasted along the chill chain or at the consumer. At the same time, a large number of people get sick every year due to spoiled food.

UK government to replace sell by/use by dates?

According to the UK minister Hilary Benn the use by date and sell by dates are old technologies that are outdated and should be replaced by other solutions or disposed of altogether.

How to Enhance Food Safety

There is a number of ways to enhance sell by and use by dates. These include better education of consumers on how to use, transport, and store fresh food products, but also by enhancing the use by and sell by dates by adding to the package smart indicators such as TTIs (Time Temperature Indicators). These show through a visible color change whether the product is still fresh. TTIs are already in use by retailers and food producers in France (Monoprix and Carrefour), Switzerland (Kneuss), and other countries in western Europe.

Codex Alimentarius

In 2003, the WHO and FAO published the Codex Alimentarius which serves as a guideline to food safety.

When heat is used in the preparation of food, it can kill or inactivate potentially harmful organisms including bacteria and viruses.

The effect will depend on temperature, cooking time, and technique used. The temperature range from 41 °F to 135 °F (5 °C to 57 °C) is the "food danger zone." Between these temperatures bacteria can grow rapidly. Under optimal conditions, *E. coli*, for example, can double in number every twenty minutes. The food may not appear any different or spoiled but can be harmful to anyone who eats it. Meat, poultry, dairy products, and other prepared food must be kept outside of the "food danger zone" to remain safe to eat. Refrigeration and freezing do not kill bacteria, but only slow their growth. When cooling hot food, it should not be left standing or in a blast chiller for more than 90 minutes.

Cutting boards are a potential breeding ground for bacteria, and can be quite hazardous unless safety precautions are taken. Plastic cutting boards are less porous than wood and have conventionally been assumed to be far less likely to harbour bacteria.

This has been debated, and some research has shown wooden boards are far better. Washing and sanitizing cutting boards is highly recommended, especially after use with raw meat, poultry, or seafood. Hot water and soap followed by a rinse with an antibacterial cleaner (dilute bleach is common in a mixture of 1 tablespoon per gallon of water, as at that dilution it is considered food safe, though some professionals choose not to use this method because they believe it could taint some foods), or a trip through a dishwasher with a "sanitize" cycle, are effective methods for reducing the risk of illness due to contaminated cooking implements.

Effects on Nutritional Content of Food

Proponents of Raw foodism argue that cooking food increases the risk of some of the detrimental effects on food or health. They point out that the cooking of vegetables and fruit containing vitamin C both elutes the vitamin into the cooking water and degrades the vitamin through oxidation. Peeling vegetables can also substantially reduce the vitamin C content, especially in the case of potatoes where most vitamin C is in the skin. However, research has also suggested that a greater proportion of nutrients present in food is absorbed from cooked foods than from uncooked foods.

Baking, grilling or broiling food, especially starchy foods, until a toasted crust is formed generates significant concentrations of acrylamide, a possible carcinogen.

Cooking dairy products may reduce a protective effect against colon cancer. Researchers at the University of Toronto suggest that ingesting uncooked or unpasteurized dairy products may reduce the risk of colorectal cancer. Mice and rats fed uncooked sucrose, casein, and beef tallow had one-third to one-fifth the incidence of microadenomas as the mice and rats fed the same ingredients cooked. This claim, however, is contentious. According to the Food and Drug Administration of the United States, health benefits claimed by raw milk advocates do not exist. "The small quantities of antibodies in milk are not absorbed in the human intestinal tract," says Barbara Ingham, Ph.D., associate professor and extension food scientist at the University of Wisconsin-Madison. "There is no scientific evidence that raw milk contains an anti-arthritis factor or that it enhances resistance to other diseases."

Several studies published since 1990 indicate that cooking muscle meat creates heterocyclic amines (HCAs), which are thought to increase cancer risk in humans. Researchers at the National Cancer Institute found that human subjects who ate beef rare or medium-rare had less than one third the risk of stomach cancer than those who ate beef medium-well or well-done.

While eating muscle meat raw may be the only way to avoid HCAs fully, the National Cancer Institute states that cooking meat below 212 °F (100 °C) creates "negligible amounts" of HCAs. Also, microwaving meat before cooking may reduce HCAs by 90%. Nitrosamines, present in processed and cooked foods, have also been noted as being carcinogenic, being linked to colon cancer.

Research has shown that grilling or barbecuing meat and fish increases levels of carcinogenic Polycyclic aromatic hydrocarbons (PAH). However, meat and fish only contribute a small proportion of dietary PAH intake - most intake comes from cereals, oils and fats. German research in 2003 showed significant benefits in reducing breast cancer risk when large amounts of raw vegetable matter are included in the diet. The authors attribute some of this effect to heat-labile phytonutrients.

Heating sugars with proteins or fats can produce Advanced glycation end products ("glycotoxins"). These have been linked to ageing and health conditions such as diabetes.

Science of Cooking

The application of scientific knowledge to cooking and gastronomy has become known as molecular gastronomy. This is a subdiscipline of food science. Important contributions have been made by scientists, chefs and authors such as Herve This (chemist), Nicholas Kurti (physicist), Peter Barham (physicist), Harold McGee (author), Shirley Corriher (biochemist, author), Heston Blumenthal (chef), Ferran Adria (chef), Robert Wolke (chemist, author) and Pierre Gagnaire (chef).

Chemical processes central to cooking include the Maillard reaction - a form of non-enzymatic browning involving an amino acid, a reducing sugar and heat.

Home-cooking vs. Factory Cooking

Although cooking has traditionally been a process carried out informally at home or around a communal fire, cooking is often, and increasingly, carried out outside the home. Bakeries were an early form of cooking outside the home, and bakeries in the past often offered the cooking of foods provided by their customers as an additional service. In the present day, factory food preparation is rapidly becoming the norm, with many "ready-to-eat" foods being prepared and cooked in factories.

"Home-cooking" may be associated with comfort food, and some commercially produced foods are presented as having been "home-cooked", regardless of their actual origin.

The term "cooking" encompasses a vast range of methods, tools, and combinations of ingredients to improve the flavour or digestibility of food. Cooking technique, known as culinary art, generally requires the selection, measurement, and combining of ingredients in an ordered procedure in an effort to achieve the desired result. Constraints on success include the variability of ingredients, ambient conditions, tools, and the skill of the individual cook. The diversity of cooking worldwide is a reflection of the myriad nutritional, aesthetic, agricultural, economic, cultural, and religious considerations that affect it.

Cooking requires applying heat to a food which usually, though not always, chemically changes the molecules, thus changing its flavour, texture, appearance, and nutritional properties. Cooking certain proteins, such as egg whites, meats, and fish, denatures the protein, causing it to firm. There is archaeological evidence of roasted foodstuffs at *Homo erectus* campsites dating from 420,000 years ago. Boiling as a means of cooking requires a container, and has been practiced at least since the 10th millennium BC with the introduction of pottery.

Cooking Equipment

There are many different types of equipment used for cooking.

Ovens are mostly hollow devices that get very hot (up to 500 °F) and are used for baking or roasting and offer a dry-heat cooking method. Different cuisines will use different types of ovens; for example, Indian culture uses a Tandoor oven, which is a cylindrical clay oven which operates at a single high temperature. Western kitchens use variable temperature convection ovens, conventional ovens, toaster ovens, or non-radiant heat ovens like the microwave oven. Classic Italian cuisine includes the use of a brick oven containing burning

Food Processing

wood. Ovens may be wood-fired, coal-fired, gas, electric, or oil-fired. Various types of cook-tops are used as well. They carry the same variations of fuel types as the ovens mentioned above. Cook-tops are used to heat vessels placed on top of the heat source, such as a sauté pan, sauce pot, frying pan, or pressure cooker. These pieces of equipment can use either a moist or dry cooking method and include methods such as steaming, simmering, boiling, and poaching for moist methods, while the dry methods include sautéing, pan frying, and deep-frying.

In addition, many cultures use grills for cooking. A grill operates with a radiant heat source from below, usually covered with a metal grid and sometimes a cover. An open pit barbecue in the American south is one example along with the American style outdoor grill fueled by wood, liquid propane, or charcoal along with soaked wood chips for smoking. A Mexican style of barbecue is called barbacoa, which involves the cooking of meats such as whole sheep over an open fire. In Argentina, an asado (Spanish for "grilled") is prepared on a grill held over an open pit or fire made upon the ground, on which a whole animal or smaller cuts are grilled.

Raw Food Preparation

Certain cultures highlight animal and vegetable foods in their raw state. Salads consisting of raw vegetables or fruits are common in many cuisines. Sashimi in Japanese cuisine consists of raw sliced fish or other meat, and sushi often incorporates raw fish or seafood. Steak tartare and salmon tartare are dishes made from diced or ground raw beef or salmon, mixed with various ingredients and served with baguettes, brioche, or frites. In Italy, carpaccio is a dish of very thinly sliced raw beef, drizzled with a vinaigrette made with olive oil. The health food movement known as raw foodism promotes a mostly vegan diet of raw fruits, vegetables, and grains prepared in various ways, including juicing, food dehydration, sprouting, and other methods of preparation that do not heat the food above 118 °F (47.8 °C).

A ceviche is a Latin American dish made with raw meat that is "cooked" from the highly acidic citric juice from lemons and limes along with other aromatics such as garlic.

Restaurants

Restaurants employ trained chefs who prepare food, and trained waitstaff to serve the customers. The term restaurant is credited to the French from the 19th century, as it relates to the restorative nature of the bouillons that were once served in them. However, the concept pre-dates the naming of these establishments, as evidence suggests commercial food preparation may have existed during the age of the city of Pompeii, and urban sales of prepared foods may have existed in China during the Song Dynasty. The coffee shops or cafés of 17th century Europe may also be considered an early version of the restaurant. In 2005, the population of the United States spent $496 billion for out-of-home dining. Expenditures by type of out-of-home dining were as follows: 40% in full-service restaurants, 37.2% in limited service restaurants (fast food), 6.6% in schools or colleges, 5.4% in bars and vending machines, 4.7% in hotels and motels, 4.0% in recreational places, and 2.2% in others, which includes military bases.

Food Manufacture

Food manufacture is the process by which food is manufactured.

History

Early food processing techniques were limited by the available food preservation, packaging and transportation. Early food processing mainly involved salting, curing, curdling, drying, pickling and smoking. An example of an early processed food product is cheese.

During the industrialisation era in the 19th century, food manufacturing arose. This development took advantage of new mass markets and emerging new technology, such as milling, preservation, packaging and labelling and transportation. It brought the advantages of prepared time-saving food to the bulk of ordinary people who did not employ domestic servants.

Industry Organization

At the start of the 21st century, a two-tier structure has arisen, with a few international food processing giants controlling a wide range of well-known food brands, and a populous number of small local or national food processing companies. Packaged foods are manufactured outside the home for purchase. This can be as simple as a butcher preparing meat, or as complex as a modern international food industry. Early food processing techniques were limited by available food preservation, packaging, and transportation. This mainly involved salting, curing, curdling, drying, pickling, fermenting, and smoking. Food manufacturing arose during the industrial revolution in the 19th century. This development took advantage of new mass markets and emerging new technology, such as milling, preservation, packaging and labelling, and transportation. It brought the advantages of pre-prepared time-saving food to the bulk of ordinary people who did not employ domestic servants.

At the start of the 21st century, a two-tier structure has arisen, with a few international food processing giants controlling a wide range of well-known food brands. There also exists a wide array of small local or national food processing companies. Advanced technologies have also come to change food manufacture. Computer-based control systems, sophisticated processing and packaging methods, and logistics and distribution advances can enhance product quality, improve food safety, and reduce costs.

Production Processes

Extrusion is the forcing or thrusting out of an object. Within most industries, extrusion is used in tandem with molds and cutting mechanisms to create uniform, identically-shaped and formed objects.

Within the food industry, contract extrusion can have many applications. It can be used in the production and packaging of cereals, noodles, cookie dough, and many other items commonly found in grocery stores. Twin-screw extruders are used to grind and mix ingredients before extruding them through a mold to the desired shape.

On a twin-screw extruder, an engine powers two large, screw-shaped devices that rotate opposite one another. The distance between the two screws determines the size to which the ingredients are ground. The ingredients can also be heated or cooled throughout the food extrusion process.

At the start of the extrusion process, raw ingredients are fed into the primary feed port. In nearly all cases, these are dry ingredients. Immediately after being fed into the machine, the ingredients are crushed and ground into ideally sized particles. In complex mixtures, an extruder may have multiple ports along its sides. These ports allow for mid-process additions of other ingredients. The more ports, the more ingredients can be added into the mixture. This allows for the gradual combination of ingredients, to ensure ideal composition and integrity of the product. Twin-screw extruders are used primarily for mixtures, but single-components can be run through the extruder to achieve the ideal consistency and final product.

Now that the ingredients have been mixed, they can be heated, cooled, dried, or added into another mixture. The ingredients or products, having been sufficiently ground and combined, move into the actual extrusion process. In this step of the process, they are forced through a die to give the product the desired shape. Think of it like pushing play dough through a mold. These die can be shaped in an incredible array of options, and can create hollow spaces as well as solid products. Once pushed through the die, the shaped product is shorn off, and the extrusion process is complete. Depending on the product, the sheering can be rapid, for when the product is small, or slow, for when more of the product must be extruded before cutting it off.

One benefit of the extrusion process is in the quality and consistency of the finished product. It also allows you to replicate the exact product endlessly, so long as you continue to feed the ingredients into the extruder.

As you may suspect, few manufacturing and packaging facilities have this capability in-house. Contract extrusion, then, is the process by which one company or purchaser utilizes an outside company to complete their extrusion for them. Depending on many elements of the business, industry and specific technical requirements, contract extrusion and outsourcing the extrusion process can save a company money, time, and manpower.

Commercial Trade

International Exports and Imports

The World Bank reported that the European Union was the top food importer in 2005, followed at a distance by the USA and Japan. Food is now traded and marketed on a global basis. The variety and availability of food is no longer restricted by the diversity of locally grown food or the limitations of the local growing season. Between 1961 and 1999, there was a 400% increase in worldwide food exports. Some countries are now economically dependent on food exports, which in some cases account for over 80% of all exports.

In 1994, over 100 countries became signatories to the Uruguay Round of the General Agreement on Tariffs and Trade in a dramatic increase in trade liberalization. This included an agreement to reduce subsidies paid to farmers, underpinned by the WTO enforcement of agricultural subsidy, tariffs, import quotas, and settlement of trade disputes that cannot be bilaterally resolved. Where trade barriers are raised on the disputed grounds of public health and safety, the WTO refer the dispute to the Codex Alimentarius Commission, which was founded in 1962 by the United Nations Food and Agriculture

Organization and the World Health Organization. Trade liberalization has greatly affected world food trade.

Marketing and Retailing

Food marketing brings together the producer and the consumer. It is the chain of activities that brings food from "farm gate to plate". The marketing of even a single food product can be a complicated process involving many producers and companies. For example, fifty-six companies are involved in making one can of chicken noodle soup. These businesses include not only chicken and vegetable processors but also the companies that transport the ingredients and those who print labels and manufacture cans. The food marketing system is the largest direct and indirect non-government employer in the United States.

In the pre-modern era, the sale of surplus food took place once a week when farmers took their wares on market day into the local village marketplace. Here food was sold to grocers for sale in their local shops for purchase by local consumers. With the onset of industrialization and the development of the food processing industry, a wider range of food could be sold and distributed in distant locations. Typically early grocery shops would be counter-based shops, in which purchasers told the shop-keeper what they wanted, so that the shop-keeper could get it for them.

In the 20th century, supermarkets were born. Supermarkets brought with them a self service approach to shopping using shopping carts, and were able to offer quality food at lower cost through economies of scale and reduced staffing costs. In the latter part of the 20th century, this has been further revolutionized by the development of vast warehouse-sized, out-of-town supermarkets, selling a wide range of food from around the world. Unlike food processors, food retailing is a two-tier market in which a small number of very large companies control a large proportion of supermarkets. The supermarket giants wield great purchasing power over farmers and processors, and strong influence over consumers. Nevertheless, less than 10% of consumer spending on food goes to farmers, with larger percentages going to advertising, transportation, and intermediate corporations.

Prices

It was reported on March 24, 2008, that consumers worldwide faced rising food prices. Reasons for this development include changes in the weather and dramatic changes in the global economy, including higher oil prices, lower food reserves, and growing consumer demand in China and India. In the long term, prices are expected to stabilize. Farmers will grow more grain for both fuel and food and eventually bring prices down. Already this is happening with wheat, with more crops to be planted in the United States, Canada, and Europe in 2009. However, the Food and Agriculture Organization projects that consumers still have to deal with more expensive food until at least 2018.

It is rare for the spikes to hit all major foods in most countries at once. Food prices rose 4% in the United States in 2007, the highest rise since 1990, and are expected to climb as much again in 2008. As of December 2007, 37 countries faced food crises, and 20 had imposed some sort of food-price controls. In China, the price of pork jumped 58% in 2007. In the 1980s and 1990s, farm subsidies and support programs allowed major grain exporting countries to hold large surpluses, which could be tapped during food shortages to keep prices down. However, new trade policies have made agricultural production much more responsive to market demands, putting global food reserves at their lowest since 1983.

Food prices are rising, wealthier Asian consumers are westernizing their diets, and farmers and nations of the third world are struggling to keep up the pace. The past five years have seen rapid growth in the contribution of Asian nations to the global fluid and powdered milk manufacturing industry, which in 2008 accounted for more than 30% of production, while China alone accounts for more than 10% of both production and consumption in the global fruit and vegetable processing and preserving industry.

The trend is similarly evident in industries such as soft drink and bottled water manufacturing, as well as global cocoa, chocolate, and sugar confectionery manufacturing, forecast to grow by 5.7% and 10.0% respectively during 2008 in response to soaring demand in Chinese and Southeast Asian markets.

Famine and Hunger

Food deprivation leads to malnutrition and ultimately starvation. This is often connected with famine, which involves the absence of food in entire communities. This can have a devastating and widespread effect on human health and mortality. Rationing is sometimes used to distribute food in times of shortage, most notably during times of war.

Starvation is a significant international problem. Approximately 815 million people are undernourished, and over 16,000 children die per day from hunger-related causes. Food deprivation is regarded as a deficit need in Maslow's hierarchy of needs and is measured using famine scales.

Food Aid

Food aid can benefit people suffering from a shortage of food. It can be used to improve peoples' lives in the short term, so that a society can increase its standard of living to the point that food aid is no longer required. Conversely, badly managed food aid can create problems by disrupting local markets, depressing crop prices, and discouraging food production. Sometimes a cycle of food aid dependence can develop. Its provision, or threatened withdrawal, is sometimes used as a political tool to influence the policies of the destination country, a strategy known as food politics. Sometimes, food aid provisions will require certain types of food be purchased from certain sellers, and food aid can be misused to enhance the markets of donor countries. International efforts to distribute food to the neediest countries are often coordinated by the World Food Programme.

Safety

Foodborne illness, commonly called "food poisoning", is caused by bacteria, toxins, viruses, parasites, and prions. Roughly 7 million people die of food poisoning each year, with about 10 times as many suffering from a non-fatal version. The two most common factors leading to cases of bacterial foodborne illness are cross-contamination of ready-to-eat food from other uncooked foods and improper temperature control. Less commonly, acute adverse reactions can also occur if chemical contamination of food occurs, for example from improper storage, or use of non-food grade soaps and disinfectants. Food can also be adulterated by a very wide range of articles (known as "foreign bodies") during farming, manufacture, cooking, packaging, distribution, or sale. These foreign bodies can include pests or their droppings, hairs, cigarette butts, wood chips, and all manner of other contaminants. It is possible for certain types of food to become contaminated if stored or presented in an unsafe container, such as a ceramic pot with lead-based glaze.

Food poisoning has been recognized as a disease since as early as Hippocrates. The sale of rancid, contaminated, or adulterated food was commonplace until the introduction of hygiene, refrigeration, and vermin controls in the 19th century. Discovery of techniques for killing bacteria using heat, and other microbiological studies by scientists such as Louis Pasteur, contributed to the modern sanitation standards that are ubiquitous in developed nations today. This was further underpinned by the work of Justus von Liebig, which led to the development of modern food storage and food preservation methods. In more recent years, a greater understanding of the causes of food-borne illnesses has led to the development of more systematic approaches such as the Hazard Analysis and Critical Control Points (HACCP), which can identify and eliminate many risks.

Recommended measures for ensuring food safety include maintaining a clean preparation area with foods of different types kept separate, ensuring an adequate cooking temperature, and refrigerating foods promptly after cooking.

Foods that spoil easily, such as meats, dairy, and seafood, must be prepared a certain way to avoid contaminating the people for whom they are prepared. As such, the general rule of thumb is that cold foods (such as dairy products) should be kept cold and hot foods (such as soup) should be kept hot until storage. Cold meats, such as chicken, that are to be cooked should not be placed at room temperature for thawing, at the risk of dangerous bacterial growth, such as *Salmonella* or *E. coli*.

Allergies

Some people have allergies or sensitivities to foods which are not problematic to most people. This occurs when a person's immune system mistakes a certain food protein for a harmful foreign agent and attacks it. About 2% of adults and 8% of children have a food allergy.

The amount of the food substance required to provoke a reaction in a particularly susceptible individual can be quite small. In some instances, traces of food in the air, too minute to be perceived through smell, have been known to provoke lethal reactions in extremely sensitive individuals.

Common food allergens are gluten, corn, shellfish (mollusks), peanuts, and soy. Allergens frequently produce symptoms such as diarrhea, rashes, bloating, vomiting, and regurgitation. The digestive complaints usually develop within half an hour of ingesting the allergen. Rarely, food allergies can lead to a medical emergency, such as anaphylactic shock, hypotension (low blood pressure), and loss of consciousness. An allergen associated with this type of reaction is peanut, although latex products can induce similar reactions. Initial treatment is with epinephrine (adrenaline), often carried by known patients in the form of an Epi-pen or Twinject.

Diet

In nutrition, diet is the sum of food consumed by a person or other organism. Dietary habits are the habitual decisions an individual or culture makes when choosing what foods to eat. With the word diet, it is often implied the use of specific intake of nutrition for health or weight-management reasons (with the two often being related). Although humans are omnivores, each culture and each person holds some food preferences or some food taboos, due to personal tastes or due to ethical reasons. Individual dietary choices may be more or less healthful. Proper nutrition requires the proper ingestion and, equally important, the absorption of vitamins, minerals, and food energy in the form of carbohydrates, proteins, and fats. Dietary habits and choices play a significant role in health and mortality, and can also define cultures and play a role in religion.

Traditional Diets

Traditional diets are those of native populations such as the Native Americans, Khoisan or Australian Aborigines. Often, to qualify for cultural cuisine, traditional diets include more organic farming and seasonal food according to food origins.

Traditional diets vary with availability of local resources, such as fish in coastal towns, eels and eggs in estuary settlements, or squash, corn and beans in farming towns, as well as with cultural and religious customs and taboos. In some cases, the crops and domestic animals that characterize a traditional diet have been replaced by modern high-yield crops, and are no longer available. The slow food movement attempts to counter this trend and to preserve traditional diets.

A recent study has suggested that traditional diets may have been more balanced than first thought. New research indicates grains were part of the diet of ancient people in Italy, Russia and the Czech Republic.

Religious and Cultural Dietary Choices

Some cultures and religions have restrictions concerning what foods are acceptable in their diet. For example, only Kosher foods are permitted by Judaism, and Halal foods by Islam.

Diet and Life Outcomes

A three-decade long study published in the British medical journal, The Lancet, found that Guatemalan men who had been well-fed soon after they were born earned almost 50% more in average salary than those who had not. The blind trial was performed by giving a high-nutrition supplement to some infants and a lower-nutrition supplement to others, with only the researchers knowing which infants received which supplements. The infants that received the high-nutrition supplement had higher average salaries as adults.

Individual Dietary Choices

Writers such as Michael Pollan and Mark Bittman urge reduced animal consumption in the developed world for improved health and reduced impact on the environment. Many people choose to forgo food from animal sources to varying degrees (flexitarianism, vegetarianism, veganism, fruitarianism) for health reasons, or issues surrounding morality, or to reduce their personal impact on the environment. Raw foodism is another contemporary trend. These diets may require tuning or supplementation to meet ordinary nutritional needs.

Diets for Weight Management

A particular diet may be chosen to seek weight loss or weight gain. Changing a subject's dietary intake, or "going on a diet", can change the energy balance and increase or decrease the amount of fat stored by the body. Some foods are specifically recommended, or even altered, for conformity to the requirements of a particular diet. These diets are often recommended in conjunction with exercise. Specific weight loss programs can be harmful to health, while others may be beneficial (and can thus be coined as healthy diets). The terms healthy diet and diet for weight management are often related, as the two promote healthy weight management.

Health

A healthy diet is one that is arrived at with the intent of improving or maintaining optimal health. This usually involves consuming nutrients by eating the appropriate amounts from all of the food groups, including an adequate amount of water. Since human nutrition is complex, a healthy diet may vary widely, and is subject to an individual's genetic make-up, environment, and health. For around 20% of the human population, lack of food and malnutrition are the main impediments to healthy eating. Conversely, people in developed countries have the opposite problem; concern is not about volume of food but appropriate choices.

Chapter 2

Enzymes in Food Processing

Food processing is a procedure in which food is prepared for consumption. People often use this term to refer specifically to making packaged foods, but technically anything which transforms raw ingredients into something else is a form of food processing, ranging from grilling vegetables in the back yard to making television dinners in a food manufacturing facility. The food processing sector employs large numbers of people, many of whom are unskilled labourers.

There are several purposes to food processing. The most basic goal is to prepare food which is palatable. This can include processing ingredients which are not safe to eat raw, flavouring foods to make them more interesting, and making dishes which comply with cultural and religions norms surrounding food, in addition to addressing issues such as allergies. Food processing is also usually intended to make food which is nutritious, and can include activities such as food fortification, in which vitamins and minerals are added to food during processing to increase the nutritional value.

Safety is also a major concern in food processing, especially industrial food processing to create packaged foods which are sold commercially. These facilities can be easily contaminated and the contamination can quickly spread, causing widespread illness. Part of making food safe includes processing it to remove any potential risks, such as bacteria in milk, in addition to maintaining strict safety procedures to reduce the risk of introducing harmful organisms during food processing.

In industrial food processing, these needs must be balanced with the need for preservation. Food which will be eaten in a few days needs to be stabilized so that it will retain texture and flavour in addition to staying safe. Other foods intended to be kept in dry storage for months or years must be specially processed so that they will not go bad. Some of the earliest forms of processing such as pickling and preserving foods echo the need to preserve foods effectively and safely which has challenged human cultures for thousands of years.

Many innovative techniques have been developed for food processing around the world to bring food to market while keeping it safe and flavourful. Certain processes are protected by patent and used only by people licensed to use the patent, such as the process behind Pringles® potato products, while others are widely used by everyone from home bakers to companies which produce diet meals on order for hospitals and institutions.

Food and feed is possibly the area where processing anchored in biological agents has the deepest roots. Despite this, process improvement or design and implementation of novel approaches has been consistently performed, and more so in recent years, where significant advances in enzyme engineering and biocatalyst design have fastened the pace of such developments. This chapter aims to provide an updated and succinct overview on the applications of enzymes in the food sector, and of progresses made, namely, within the scope of tapping for more efficient biocatalysts, through screening, structural modification, and immobilization of enzymes.

Targeted improvements aim at enzymes with enhanced thermal and operational stability, improved specific activity, modification of pH-activity profiles, and increased product specificity, among others. This has been mostly achieved through protein engineering and enzyme immobilization, along with improvements in screening. The latter has been considerably improved due to the implementation of high-throughput techniques, and due to developments in protein expression and microbial cell culture. Expanding screening to relatively unexplored environments (marine, temperature extreme environments) has also contributed to the identification and development of more efficient biocatalysts. Technological aspects are considered, but economic aspects are also briefly addressed.

Enzymes in Food Processing: A Condensed Overview on Strategies for Better Biocatalysts

Food processing through the use of biological agents is historically a well-established approach. The earliest applications go back to 6,000 BC or earlier, with the brewing of beer, bread baking, and cheese and wine making, whereas the first purposeful microbial oxidation dates from 2,000 BC, with vinegar production. Coming to modern days, in the late XIX, century Christian Hansen reported the use of rennet (a mixture of chymosin and pepsin) for cheese making, and production of bacterial amylases was started at Takamine (latter to become part of Genencor).

Pectinases were used for juice clarification in the 1930s, and for a short period during World War II, invertase was also used for the production of invert sugar syrup in a process that pioneered the use of immobilized enzymes in the sugar industry. Still, the large-scale application of enzymes only became really established in the 1960s, when the traditional acid hydrolysis of starch was replaced by an approach based in the use of amylases and amyloglucosidases (glucoamylases), a cocktail that some years latter would include glucose (xylose) isomerase. From then on, the trend for the design and implementation of processes and production of goods anchored in the use of enzymes has steadily increased. Enzymes are currently among the well established products in biotechnology, from US $1.3 billion in 2002 to US $4 billion in 2007; it is expected to have reached US $5.1 billion in a rough 2009 year, and is anticipated to reach $7 billion by 2013. In the overall, this pattern corresponds to a rise in global demand slightly exceeding 6% yearly. Part of this market is ascribed to enzymes used in large-scale applications, among them are those used in food and feed applications.

These include enzymes used in baking, beverages and brewing, dairy, dietary supplements, as well as fats and oils, and they have typically been dominating one, only bested by the segment assigned to technical enzymes. The latter includes enzymes in the detergent, personal care, leather, textile and pulp, and paper industries. A recent survey on world sales of enzymes ascribes 31% for food enzymes, 6% for feed enzymes and the remaining for technical enzymes. A relatively large number of companies are involved in enzyme manufacture, but major players are located in Europe, USA and Japan. Denmark is dominating, with Novozymes (45%) and Danisco (17%), moreover after the latter taking over Genencor (USA), with DSM (The Netherlands) and BASF (Germany) lagging behind, with 5% and 4%. The pace of development in emerging markets is suggestive that companies from India and China can join this restricted party in a very near future.

Relevant Enzymes: Tapping for Improved Biocatalysts

General Aspects and the Screening Approach

Roughly all classes of enzymes have an application within the food and feed area, but hydrolases are possibly the prevalent one. The widespread use of enzymes for food and feed processing is easily understandable, given their unsurpassed specificity, ability to operate under mild conditions of pH, temperature and pressure while displaying high activity and turnover numbers, and high biodegradability. Enzymes are furthermore generally considered a natural product. The whole contributes for developing sustainable and environmentally friendly processes, since there is a low amount of by-products, hence reducing the need for complex downstream process operations, and the energy requirements are relatively low. Life-cycle assessment (LCA) has confirmed, that within the range of given practical case studies, including food and feed processing, the implementation of enzyme-based technology has a positive impact on the environment. LCA is a methodology used to compare the environmental impact of alternative production technologies while providing the same user benefits.

Some of the broad generalizations on the limitations of enzymes for application as biocatalysts in commercial scale, namely, their high cost, low productivity and stability, and narrow range of substrates, have been rebutted. Aiming at improving the performance of biocatalysts for food and feed applications, particular care has been given to increasing thermal stability, enhancing the range of pH with catalytic activity and decreasing metal ions requirements, as well as to overcoming the susceptibility to typical inhibitory molecules.

Some examples of strategies taken to improve the performance of relevant enzymes for food and feed. Along with these different strategies focused on the enzyme molecule (namely, protein engineering, enzyme immobilization), the developments in recombinant DNA technology that occurred in the 1980s also had a huge impact on the application of enzymes in food and feed.

By allowing gene cloning in microorganisms compatible with industrial requirements, this methodology enabled cost-feasible production of enzymes that were naturally produced in conditions that prevented large-scale application (namely, enzymes from plant or animal cells, such as transglutaminase or even slow-growing microorganisms). When successfully implemented, the undertaken approaches allow: (a) continuous operations at relatively high temperatures; (b) eased implementation of enzyme cascade, given the reduced need for processing the reaction media (pH adjustments; metal ion removal/addition) throughout the intermediate steps of a multistep biotransformation (namely, starch to high fructose syrup); and (c) the use of raw substrates, preferably as high-concentrated solutions, hence cutting back in costs related to upstream processing and increasing productivity. Methodologies with a high level of parallelization, anchored in computer-monitored microtiter plates equipped with optic fibers and temperature control have also been developed. These provide high-throughput capability for a speedy and detailed characterization of the performance of enzymes. Particular focus was given to the prediction of the long-term stability of enzymes under moderate conditions using short-term runs (up to 3 hours).

One of the methodologies to obtain improved biocatalyst relies on in-vitro modifications, which will be addressed latter in this paper; another approach relies on screening efforts, which has been consistently undertaken, as summarized recently. Some focus is given to extremophiles, particularly thermophiles, since operation at high temperatures (roughly above 45–50°C) minimizes the risk of microbial contamination, a particularly delicate matter under continuous operation.

Furthermore, the extension of some reactions in relevant food applications is favored at relatively high temperatures (namely, isomerization of glucose to fructose), although care should be taken to avoid an operational environment that may lead by-product formation (namely, Maillard reactions). Examples of screened enzymes include the isolation of amylases, with some of them being calcium independent; amylopullulanases; fructosyltransferases; glucoamylases; glucose (xylose) isomerases; glucosidases; inulinases; levansucrases; pullulanases; and xylanases. Other examples of these enzymes, with some of which able to retain stability under temperatures of 90°C or higher, were reviewed by Gomes and Steiner. The majority of enzymes used in food and feed processing is of terrestrial microbial origin, and screening-efforts for isolation of promising enzyme-producing strains have accordingly been performed in such background. From some years now, marine environment has also been tapped as a source for useful enzymes from either microbial or higher organisms origin.

This latter environment has allowed the isolation of some promising biocatalysts, such as the heat-stable invertase/inulinase from Thermotoga neapolitana DSM 4359 or inulinase from Cryptococcus aureus, amylolytic enzymes, glucosidases and proteases from severalgenera, esterase from Vibrio fischeri, and glycosyl hydrolases. Other examples of useful enzymes for food and feed, but isolated from higher organisms, are given in Operation at low temperatures is also welcome since it also reduces the risk of microbial contamination, enables some processes to be carried out with minimal deterioration of the raw material. These include protein processing, such as cheese maturing and milk coagulation with proteases; milk processing with lactase for lactose-free milk; clarification of fruit juices with pectinases to produce clear juice; or production of oligosaccharides.

Since extremophiles are often difficult to grow under typical laboratory conditions if not nonculturable at all, different approaches have been developed in order to assess the potential of enzymes from such microorganisms. One approach relies on the generation and screening of target genes from DNA libraries, which can be obtained from mixed microbial population from environmental samples.

Recombinant microorganisms can then be obtained using mesophiles as hosts where the genes of interest from extremophiles have been expressed. In order to screen the huge number of DNA-libraries typically generated for the intended property, high-throughput methods have been implemented. These methods are also widely used when protein engineering is carried out. This will be addressed in the following section.

Several enzymes (namely, α-amylases; pullulanases) currently used in food processing, namely, in starch hydrolysis, are actually produced by recombinant microorganisms. Despite some complexity in the implementation of their use in large-scale applications, partly resulting from lack of uniformity in the US and EU legislation, quite a few enzyme preparations have been accepted for industrial use.

Improving Biocatalysts: Beyond Screening

Taking advantage of the knowledge gathered on molecular biology, high-throughput processing, and computer-assisted design of proteins, in-vitro improvement of biocatalysts have been consistently implemented. Some of the research efforts in this area has focused on the biochemical and molecular mechanisms underlying the stability of enzymes from extremophiles. Such knowledge is also particularly useful for protein engineering of known enzymes, aiming at enhancing stability without compromising catalytic activity. Enhancing the stability of enzymes is of paramount importance when implementation of industrial processes is foreseen, since it allows for reducing the amount of enzyme used in the process.

Given that thermostability is determined by a series of short- and long-range interactions, it can be improved by several substitutions of amino acids in a single mutant, where the combination of each individual effect is usually roughly additive. The targeted improvements have not been restricted to thermostability, but they have also addressed other features, such as broadening the range of pH where the enzyme is active, or lessening the temperature of operation while retaining high activity.

Two methodologies can be used for protein engineering. The first is directed evolution of enzymes, through random mutagenesis and recombination, where the environmental adaptation is reproduced in-vitro in a much hastened timescale, towards the optimization of the intended property. In order to control the pathway of the process, either a screening test for the assessed feature is performed after each round of modification, or selective pressure is applied.

This methodology, which allows for a high throughput, has been extensively applied, aiming for more efficient biocatalysts. Some relevant examples in the area of food and feed processing include the following.

The first is the enhancement of the activity of the hyperthermostable glucose (xylose) isomerase from Thermotoga neapolitana at relatively low temperature and pH, without decay in thermostability. The enzyme from the parent strain is highly active at 97°C, but it retains only 10% of its activity at 60°C, and requires neutral pH for optimal activity. This pattern is often reported when glucose isomerases from hyperthermophilic strains operate in mesophilic environments.

Large-scale glucose isomerization is carried out at 55–60°C and slightly alkaline pH. This set of conditions results from the optimal range of pH (typically 7.0 to 9.0) and temperature (60 to 80°C) for glucose isomerization displayed by most of the glucose isomerases used, combined with process boundary conditions. The latter result from by-product and color formation occurring when the reaction is carried out at alkaline pH and high temperatures. There is therefore interest in selecting an enzyme able to operate efficiently at temperatures close to those currently used but at a lower pH.

The mutant glucose isomerase 1F1 obtained by Sriprapundh and coworkers displayed a roughly 5-fold higher activity at 60°C and pH 5.5, when compared with the parent T. neapolitana isomerase, and was more thermostable than the wild type isomerase. The activation energy required by the triple 1F1 mutant (V185T/L282P/F186S) was roughly half of the wild-type, hence allowing for high activity at relatively low temperatures. The encouraging results obtained suggest the soundness of the approach to obtain a mutant glucose isomerase competitive with those currently used, while being able to operate in a slightly acidic environment and 60°C. The second is the enhancement of the thermostability of the maltogenic amylase from Thermus sp. IM6501, of the amylosucrase from Neisseria polysaccharea, of the glucoamylase from Aspergillus niger, of a phytase from Escherichia coli, and of a xylanase from Bacillus subtilis.

Amylases and glucoamylases are enzymes used in starch processing, which involves temperatures typically in excess of 60°C; hence, improving thermal stability without decreasing enzyme activity is of relevance. Starch liquefaction is performed at 105°C in the presence of α-amylase, upon which the effluent reaction stream has to be cooled to 60°C, so that glucoamylases can be used.

In order to avoid, or at least minimize, the cooling step, thermostable glucoamylases are aimed at. Wang and coworkers obtained a multiply-mutated enzyme (N20C, A27C, S30P, T62A, S119P, G137A, T290A, H391Y), which displayed a 5.12 kJmol-1 increase in the free energy of thermal inactivation, when compared to the wild type, thus resulting in the enhanced thermal stability of the mutant.

Furthermore specific activities and catalytic efficiencies remained unaltered, when mutant and wild type were compared. Kim and coworkers obtained also a multiply-mutated amylase (R26Q, S169N, I333V, M375T, A398V, Q411L, P453L) which displayed an optimal reaction temperature 15°C higher than that of the wild-type and a half-life of roughly 170 min at 80°C, a temperature at which the wild-type ThMA was fully inactivated in less than 1 minute.

However, one of the mutations most accountable for enhanced thermal stability, M375T, close to the active site, also led to a 23% decrease in specific activity, as compared to the wild type. The amylosucrase engineered by Emond and coworkers was a double mutant (R20C/A451T), displaying a 10-fold increase in the half-life at 50°C compared to the wild-type enzyme. Actually, the mutant was claimed to be the only amylosucrase usable at 50°C. At the latter temperature, the mutant enabled the synthesis of amylose chains twice as long as those obtained by the wild-type enzyme at 30°C, for sucrose concentrations of 600 mM.

The mutant thus allowed for a process with increased yield in amylose chains, lower risk of contamination, enhanced substrate and product solubility and overall productivity. Phytases are added to animal feeds to improve phosphorus nutrition and to reduce phosphorus excretion, by promoting the hydrolysis of phytate into myoinositol and inorganic phosphate. Thermal stable enzymes are needed, since feed pelleting is carried out at high temperature (60 to 80°C).

Phytases produced by thermophiles do not provide a suitable approach, since they have low activity at the physiological temperature of animals. E. coli phytases, which are appealing to industrial application, due to the acidic pH optimum, specificity phytate, and resistance to pepsin digestion, were thus engineered in order to improve their thermal stability, without compromising the kinetic parameters. As a result, mutants were obtained, with roughly 20% increased thermostability at 80°C improved overall catalytic efficiency (kcat, turnover number/KM, Michaelis constant) within 50 to 150%, as compared to the wild type. No significant changes in the pH activity profile were observed, but for some mutants, containing a K46E substitution, that displayed a decrease in activity at pH 5.0.

Xylanases catalyze the cleavage of β1,4 bonds in xylan polymers. Accordingly, these enzymes can be used in dough making, in baking, in brewing and in animal feed compositions. When the latter contain cereals (namely, barley, maize, rye or wheat), or cereal by-products, xylanases improve the break-down of plant cell walls, which favours the ingestion of plant nutrients by the animals and consequently enhances feed consumption and growth rate.

Furthermore, the use of xylanases decreases the viscosity of xylan-containing feeds. As referred for phytases, the formulation of commercial feed often involves steps at high temperatures. Xylanases added to the formulations hence have to withstand these conditions, while they are to display high activity at about 40°C, which is the temperature in the intestine of animals.

However, most xylanases are inactive at temperatures exceeding 60°C, hence the need for enhancing thermal stability. Miyazaki and coworkers obtained a triple-mutant xylanase (Q7H, N8F, and S179C) which retained full activity for 2 hours at 60°C, whereas the wild-type enzyme was inactivated within 5 minutes under the same conditions.

The mutation also led to a 10°C increase in the optimal temperature for reaction and enhanced activity at higher temperatures, albeit at the cost of decreased activity at lower temperatures, as compared to the wild-type enzyme. Third is the enhancement of the activity of the amylosucrase from Neisseria polysaccharea.

Amylosucrases can be used for the modification or synthesis of amylose-type polymers from sucrose, but their industrial application is somehow thwarted by the low catalytic efficiency on sucrose and by side reactions leading to the formation of sucrose isomers. Van der Veen and co-works engineered mutant enzymes through error-prone

PCR that displayed increases in activity up to 5-fold and in overall catalytic efficiency up to 2-fold, when compared to the wild-type enzyme.

Furthermore, the mutants were able to produce amylose polymers from 10 mM sucrose on, unlike the wild-type enzyme. Their work provides an illustrative example on the use of random mutagenesis and recombination for the enhancement of the catalytic properties of enzymes with application on food and feed. Another example was provided by Tian and coworkers who engineered a phytase from Aspergillus niger 113 through gene shuffling, to obtain mutants with enhanced catalytic properties. Hence, K41E and E121F substitutions allowed for increases in the specific activity of 2.5- and 3.1-fold, and of affinity for sodium phytate, as expressed by decreases in KM of roughly 35% and 25%, as compared to the wild-type enzyme. Furthermore, the overall catalytic efficiency of the mutants increased 1.4- and 1.6-fold as compared to the wild type.

Other examples can be found elsewhere. The second methodology underlines that rational pinpoint modifications in one or more amino acids are made, where these changes are predicted to bring along the envisaged improvement in the targeted enzyme function.

The alterations promoted are performed based on the growing knowledge on the structure and functions of enzyme. Information on this matter mostly comes from bioinformatics, which provides data on amino-acid propensities and on protein sequences. Adequate processing of the data enable the output of generalized rules predicting the effect of mutations on enzyme properties. Also used are molecular potential functions, which, once implemented, enable the prediction of the effect of mutations in enzyme structure.

Computational tools used for enzyme engineering have been recently reviewed. Enzyme engineering through molecular simulations requires structural data from the native enzyme, which can be preferably obtained from crystallography or NMR. Otherwise a model is built based on known enzyme structures with homologous sequences. Computational methods are also welcome in directed evolution, as a tool to better lead the random mutagenesis.

Ultimately this approach is put into practice by producing a site-directed mutant, where selected amino acids are replaced with those suggested from the outcome of modelling. Some relevant examples of this strategy in the area of food and feed processing are given. These mostly aim to improve thermal stability and/or catalytic efficiency and/or to modify the range of pH/temperature where the enzyme is active—goals that were already referred to when examples of enzyme modifications using random mutagenesis were addressed.

The first example underlines the enhancement of the thermostability of the recombinant glucose (xylose) isomerase from Actinoplanes missouriensis and of glucose (xylose) isomerase from Streptomyces diastaticus; of amylases from Bacillus spp. and of glucoamylase from Aspergillus awamori. The mutant isomerase from A. missouriensis displayed an enhanced thermal stability, alongside with improved stability at different pH, as compared with the original enzyme, with no changes in catalytic properties. The double mutant isomerase (G138P, G247D) displayed a 2.5-fold increase in half-life, and additionally a 45% increase in the specific activity, when compared to the wild type.

Such features were ascribed to increased molecular rigidity due to the introduction of a proline in the turn of a random coil. Multiply-mutated amylases obtained by Declerck and coworkers displayed considered enhanced thermal stability. Based on the temperature at which amylase initial activity is reduced by 50% for a 10-minute incubation, this parameter went as high as 106°C, as compared to 83°C for the wild-type strain. Furthermore, the thermal stabilization was not accompanied by a decrease in the catalytic activity.

The work by Lin and coworkers on amylase mutants from Bacillus sp. strain TS-23 highlighted the relevance of E219 for the thermal stability of the enzyme. The mutated glucoamylases engineered by Liu and Wang allowed to establish the role of several intermolecular interactions in thermal stability of these enzymes. Thermostable enzymes were obtained through the introduction of disulfide bonds in highly flexible region in the polypeptide chain of the enzyme, as well as by the introduction of more hydrophobic residues-stabilized α-helices. Data gathered also showed that care had to be taken not to disrupt the hydrogen bond and salt linkage network in the catalytic center as a result of mutagenesis, for this could lead to a decrease in the specific activity and overall catalytic efficiency.

The second example underlines the enhancement of the pH-activity profile and of the thermostability of phytase from A. niger. This was achieved by combining several individual mutations that allowed for mutants that were quite active at pH 3.5. Efficient operation in the stomach of simple-stomached animals where phytate hydrolysis mostly occurs at a pH around 3.5, and the wild type was ineffective, was thus enabled. Furthermore, the hydrolytic activity of the mutants at pH 3.5 exceeded in roughly 1.5-fold that of the parent one at pH 5.5, which was the optimum of the latter. Mutants also retained higher residual activity after incubation within 70 to 100°C, as compared to the wild type.

The work demonstrates that cumulative improvements in pH activity and thermostability through mutation are compatible in this phytase. The third example underlines the modification of the temperature- and pH activity profile of the L-arabinose isomerase from Bacillus stearothermophilus US100. L-Arabinose isomerases catalyze the conversion of L-arabinose to L-ribulose in-vivo, but in-vitro they also isomerize D-galactose into D-tagatose. The latter keto-hexose is being used as a low-calorie bulk sweetener, since its taste and sweetness are roughly equivalent to sucrose, but the caloric value is only 30% of that of sucrose.

Although several thermostable L-arabinose isomerases have been isolated and characterized, most of these display an alkaline pH optimum. For industrial application this presents the same drawbacks of by-product and color formation referred to when the random mutation of glucose isomerases was addressed.

Hence, again arises the need for enzymes able to isomerize L-arabinose in an acidic environment and at relatively low temperature, 60 to 70°C. Operation within the latter temperature range also rules away the use of divalent ions, which stabilize isomerases at high temperatures. Rhimi and coworkers engineered two individual mutants, harbouring each N175H and Q268K mutations.

These led to broader optimal temperature range within 50 to 65°C and to enhanced stability in acidic media, respectively, when compared to the wild type.

An engineered double mutant, harbouring both modifications, displayed optimal activity within a pH range of 6.0 to 7.0 and a temperature range within 50–65°C. Such set of operational conditions matches the targeted goals and again shows that the basis for pH-activity profile and thermostability in L-arabinose isomerase are quite independent and compatible.

Cumulative enhancements in both properties in the same enzyme were thus possible. A similar pattern was also observed in the previous example dedicated to a mutant phytase. The fourth example underlines the modification of the product profile of inulosucrase from Lactobacillus reuteri and from B. subtilis. Inulosucrases are used to synthesize fructooligosaccharides or fructan polymer from sucrose. The transglycosylation catalyzed by the inulosucrase from L. reuteri leads to a wide range of fructooligosaccharides alongside with minor amounts of an inulin polymer.

In order to minimize the dispersion in the products obtained, mutants R423K and W271N were obtained, which allowed the synthesis of a significant amount of polymer and a lower amount of oligosaccharide, without significantly affecting the catalytic activity, when compared with the wild type. The data gathered showed that the "1 subsite in the inulosucrase from L. reuteri has a key role in the determination of the size of the products obtained.

Ortiz-Soto and coworkers also showed that the product profile of transfructosylation reactions could be adequately tuned through modification of target residues of an inulosucrase from B. subtilis. These authors established the effect of mutations on the reaction specificity (hydrolysis/transfructosylation), molecular weight and acceptor specificity.

For example, engineered mutants R360S, Y429N and R433A only synthesized oligosaccharides, whereas the wild type synthesized levan, since the former are more hydrolytic. On the other hand these mutations reduced the affinity for sucrose, and thermal stability, when compared to the wild type.

The fifth example underlines the enhancement of the product profile of cyclodextrin glycosyltransferases (CGTase) from differentgenera. These enzymes promote the production of cyclodextrins, linked oligosaccharides form starch, through an intramolecular transglycosylation reaction. In the process, a starch oligosaccharide is cleaved and cleaved and the resulting reducing-end sugar is transferred to the non-reducing-end sugar of the same chain.

The resulting cyclodextrin may consist of six, seven or eight, which are accordingly termed α, β, or γ-cyclodextrin, respectively. Given their ability to form inclusion complexes with small hydrophobic molecules, they are of interest for both industrial and research applications.

Wild-type CGTases typically produce a mixture of the three cyclodextrins when incubated with starch. The purification of a given cyclodextrin from the reaction mixture requires several additional steps, including selective complexation with organic solvents, which may prove restrictive for cyclodextrin applications involving human consumption.

There is therefore a clear interest in obtaining a mutant CGTase capable of producing a particular type of cyclodextrin in a high rate. Van der Veen and coworkers engineered a double-mutant (Y89D/S146P) of CGTase from Bacillus circulans which displayed a 2-fold increase in the production of α-cyclodextrin and a marked decrease in β-cyclodextrin when compared to the wild type.

From the data gathered, the authors suggested that hydrogen bonds (S146) and hydrophobic interactions (Y89), are likely to play a key role in to the size of cyclodextrin products formed, and that changes in sugar-binding subsites "3 and "7 may result in mutant CGTases with altered product specificity.

Li and coworkers were also able to obtain CGTase mutants from Paenibacillus macerans strain JFB05-01 with increased specificity for α-cyclodextrin, through mutations at subsite "3. In particular, double mutant D372K/Y89R displayed a 1.5-fold increase in the production of β-cyclodextrin, and a significant (roughly 45%) decrease in the production of g-cyclodextrin when compared to the wild-type enzyme.

The two methods are not mutually exclusive and methodologies for engineering of enzymes can assemble both strategies.

Upon identification of the most adequate enzyme, this can be formulated adequately for better process integration. One of the most widely considered approaches for such formulation is enzyme immobilization.

Immobilization

There are several issues that can be lined up to sustain enzyme immobilization. It allows for high-enzyme load with high activity within the bioreactor, hence leading to high-volumetric productivities; it enables the control of the extension of the reaction; downstream process is simplified, since biocatalyst is easily recovered and reused; the product stream is clear from biocatalyst; continuous operation (or batch operation on a drain-and-fill basis) and process automation is possible; and substrate inhibition can be minimized.

Along with this, immobilization prevents denaturation by autolysis or organic solvents, and can bring along thermal, operational and storage stabilization, provided that immobilization is adequately designed. Immobilization has some intrinsic drawbacks, namely, mass transfer limitations, loss of activity during immobilization procedures, particularly due to chemical interaction or steric blocking of the active site; the possibility of enzyme leakage during operation; risk of support deterioration under operational conditions, due to mechanical or chemical stress; and a (still) relative empirical methodology, which may hamper scale up.

Economical issues are furthermore to be taken into consideration when commercial processes are envisaged, although immobilization can prove critical for economic viability if costly enzymes are used.

Still, the cost of the support, immobilization procedure and processing the biocatalyst once exhausted, up- and downstream processing of the bioconversion systems, and sanitation requirements have to be taken into consideration.

In the overall, the enhanced stability allowing for consecutive reuse leads to high specific productivity (massproduct-1 massbiocatalyst-1), which influences biocatalyst-related production costs. A typical example is the output of immobilized glucose isomerase, allowing for 12,000–15,000kg of dry-product high-fructose corn syrup (containing 42% fructose) per kilogram of biocatalyst, throughout the operational lifetime of the biocatalyst.

Increased thermal stability, allowing for routine reactor operation above 60°C minimizes the risks of microbial growth, hence leading to lower risks of microbial growth and to less demanding sanitation requirements, since cleaning needs of the reactor are less frequent. A rule of thumb suggesting that the enzyme costs should be a few percent of the total production costs has been established.

The half-life of the bioreactor is also a critical issue when evaluating the economical feasibility of a bioconversion process, longer half-lives favouring process economics. Examples of commercial bioreactors depict half-lives of several months to years, and the same packing can work throughout some months to years.

Among this group, are immobilized enzyme reactors packed with glucose isomerase for the production of high-fructose corn syrup; lactase for lactose hydrolysis, for the production of whey hydrolysates and for the production of tagatose; aminoacylase for the production of amino acids; isomaltulose synthase for the production of isomaltulose; invertase for the production of inverted sugar syrup; lipases for the interesterification of edible oils, ultimately targeted at the production of trans-free fat, of cocoa butter equivalents, and of modified triacylglycerols; and g-fructofuranosidase for the production of fructooligosaccharides. On the other hand, despite the technical advantages of immobilization, the large-scale liquefaction of starch to dextrins by β-amylases is performed by free enzymes, given the low cost of the enzyme.

Immobilization can be performed by several methods, namely, entrapment/microencapsulation, binding to a solid carrier, and cross-linking of enzyme aggregates, resulting in carrier-free macromolecules. The latter presents an alternative to carrier-bound enzymes, since these introduce a large portion of noncatalytic material.

This can account to about 90% to more than 99% of the total mass of the biocatalysts, resulting in low space-time yields and productivities, and often leads to the loss of more than 50% native activity, which is particularly noticeable at high enzyme loadings.

A typical example of the patterns suggested by data was observed by Abdel-Naby when evaluating the immobilization of β-amylase through different methods. Details on the different methods, as well as some illustrative examples of their applications, are given hereafter.

Entrapment (micro) encapsulation, where the enzyme is contained within a given structure. This can be: a polymer network of an organic polymer or a sol-gel; a membrane device such as a hollow fiber or a microcapsule; or a (reverse) micelle.

Apart from the hollow fiber, the whole process of immobilization is performed in-situ. The polymeric network is formed in the presence of the enzyme, leading to supports that are often referred to as beads or capsules.

Still, the latter term could preferably be used when the core and the boundary layer(s) are made of different materials, namely, alginate and poly-L-lysine. Although direct contact with an adverse environment is prevented, mass transfer limitations may be relevant, enzyme loading is relatively low, and leakage, particularly of smaller enzymes from hydrogels (namely, alginate, gelatin), may occur.

This may be minimized by previously cross-linking the enzyme with multi-functional agent (namely, glutaraldehyde) or by promoting cross-linkage of the matrix after the entrapment. The use of LentiKats, a polyvinyl-alcohol-based support in lens-shaped form, has been used for several applications in carbohydrate processing. Among these are the synthesis of oligosaccharides with dextransucrase, maltodextrin hydrolysis with glucoamylase, lactose hydrolysis with lactase, and production of invert sugar syrup with invertase.

In these processes the biocatalyst could be effectively reused or operated in a continuous manner. Methodologies for large scale production of these supports have been implemented.

Flavourzyme, (a fungal protease/peptidase complex) entrapped in calcium alginate, k-carragenan, gellan, and higher melting-fat fraction of milk fat, was effectively used in cheese ripening, in order to speed up the process, while avoiding the problems associated with the use of free enzyme. These include deficient enzyme distribution, reduced yield and poor-quality cheese, partly ascribed to excessive proteolysis and whey contamination.

The enzyme complex is released in a controlled manner due to pressure applied during cheese curd. Calcium alginate beads were also used to immobilize glucose isomerase and β-amylase for starch hydrolysis to whey.

In the latter work, the authors observed that increasing the concentration of $CaCl_2$ and of sodium alginate to 4% and 3%, respectively, enzyme leakage was minimized (a common drawback of hydrogels) while allowing for high activity and stability.

This effect was also observed in a previous work where alginate-entrapped inulinase was used for sucrose hydrolysis. The stability of an amylase immobilized biocatalyst was further enhanced with the addition of 1% silica gel to the alginate prior to gelation, as reflected by the use of the biocatalyst in 20 cycles of operation, while retaining more than 90% of the initial efficiency.

Several enzymes, namely, chymosin, cyprosin, lactase, Neutrase, trypsin, have also been immobilized in liposomes. In a particularly favored technique immobilization of enzymes in liposomes, known as dehydration-rehydration vesicles (DRVs), small (diameters usually below 50nm) unilamellar vesicles (SUVs) is prepared in distilled water and mixed with an aqueous solution of the enzyme to be encapsulated. The resulting vesicle suspension is then dehydrated under freeze drying or equivalent method.

Upon rehydration, the resulting DRVs are multilamellar and larger (from 200nm to a little above 1000nm) than the original SUVs, and can capture solute molecules. Recent work in this particular application has used lactase as enzyme model and has focused on the optimization and characterization of the liposome-based immobilized system. If liposome-based biocatalysts are used in a process under continuous operation, biocatalyst separation has to be integrated (namely, using an ultra-filtration membrane).

In a different concept, based in batch mode, liposome-encapsulated lactase was incorporated in milk. After ingestion, the vesicles are disrupted in the stomach by the presence of bile salts, allowing in-situ degradation of lactose. Cocktails of enzymes, namely, Flavourzyme, bacterial proteases and Palatase M (a commercial lipase preparation), were immobilized in liposomes and successfully used to speed up cheddar cheese ripening.

Encapsulation in lipid vesicles has been proved a mild method, providing high protection against proteolysis. There is however some lack of consensus on the feasibility of its application on large scale, as well as on the effectiveness of the methodology for controlled release of enzymes. Containment within an ultra-filtration (UF) membrane allows the enzyme to perform in a fully fluid environment; hence, with little loss (if any) of catalytic activity.

However, the membrane still presents a boundary for overall mass transfer of substrate/products and enzyme molecules are prone to interact with the membrane material. This feature is enhanced along with the hydrophobicity of the membrane, hence immobilization in membrane devices may have some adsorptive nature, a feature that will be addressed in (ii).

Besides, regular replacement of the membrane may be required. Enzyme containment by a membrane has been used for the continuous production of galactooligosaccharides from lactose.

The reaction, with up to 80% lactose conversion out of a substrate concentration of 250 gL-1, was carried out in a perfectly mixed reactor and enzyme was recovered in a 10 kDa nominal molecular weight cutoff. The resulting product presented some similarities to the commercially available Vivinal prebiotic. Within the same methodology, a hollow-fiber module was used to contain lactase, in order to carry out lactose hydrolysis in continuous operation. A conversion rate close to 95% in skim milk was observed for an initial substrate concentration close to 40 gL-1.

Binding to a solid carrier, where enzyme-support interaction can be of covalent, ionic, or physical nature. The latter comprehends hydrophobic and van der Waals interactions. These are of weak nature and easily allow for enzyme leakage from the support, namely, after environmental shifts in pH, ionic strength, temperature or even as a result of flow rate or abrasion.

On the other hand, desorption can be turned into an advantage if performed under a controlled manner, since it enables the expedite removal of spent enzyme and its replacement with fresh enzyme. A recent paper by Gopinath andSugunanillustrates the increased trend for leakage when adsorption is compared with covalent binding, using β-amylase as model enzyme.

Curiously, the first reported application of enzyme immobilization was of invertase onto activated charcoal. Recently invertase was immobilized in different types of sawdust, aiming at its application for sucrose hydrolysis.

When wood shavings were used as support, the immobilized invertase retained 90% of the original activity after 20 cycles of 15 minutes, each under consecutive batch operation; and it retained 65% of the original activity after 10 hours of continuous operational regime in a column reactor. Anther example is the immobilization of pectinase in egg shell for the preparation of low-methoxyl pectin. The immobilized biocatalyst could be reused for 32 times at 30°C, and it was used in a fluidized-bed reactor, operated at an optimum flow rate of 5 mL h-1 and 35°C.

Other examples are the surface immobilizations of β-amylase on alumina and in zirconia. Covalent binding is the strongest form of enzyme linking to a solid support. It involves chemically reactive sites of the protein such as amino groups, carboxyl groups, and phenol residues of tyrosine; sulfhydryl groups; or the imidazole group of histidine.

The binding can be carried out by several methods; among them are amide bond formation, alkylation and arylation, or UGI reaction. However, this often brings along loss of activity during the process of immobilization, due to support binding to critical residues for enzyme activity, and steric hindrance, among others.

Examples include the immobilization of β-amylase and of levansucrase on glutaraldehyde-treated chitosan beads, through the glutaraldehyde reaction between the free amino groups of chitosan and the enzyme molecule; the immobilization of pectinase onto Amberlite IRA900 Cl through glutaraldehyde cross-linking; glucoamylase onto dried oxidized bagasse, onto polyglutaraldehydeactivated gelatin, or onto macroporous copolymer of ethylene glycol dimethacrylate and glycidyl methacrylate through the carbohydrate moiety of the enzyme; glucoamylase or invertase immobilized onto montmorillonite K-10 activated with aminopropyltriethoxysilane and glutaraldehyde; and invertase immobilized on nylon-6 microbeads, previously activated with glutaraldehyde and using PEI as spacer; on polyurethane treated with hydrochloric acid, polyethylenimine and glutaraldehyde; on poly(styrene-2-hydroxyethyl methacrylate) microbeads activated with epichlorohydrin; or on poly(hydroxyethyl methacrylate)/glycidyl methacrylate films.

Within this methodology for immobilization, highlight should be given to the introduction of commercial supports (namely, Eupergit, Sepabeads) with a high density of epoxide functional groups aimed at multipoint attachment, typically with the g-amino group of lysine, to confer high rigidity to the enzyme molecule, hence enhancing stabilization.

This methodology has been used for lactase immobilization in magnetic poly (GMA-MMA), formed from monomers of glycidylmethacrylate and ethylmethacrylate, and cross-linked with ethyleneglycol dimethacrylate; for the immobilization of cyclodextrin glycosyltransferases to glyoxylagarose supports for the production of cyclodextrins; or for the immobilization of dextransucrase on Eupergit C. Ionic binding to a carrier involves interaction of negatively or positively charged groups of the carrier with charged amino-acid residues on the enzyme molecules.

Ionic interaction may be favored if enzyme leakage is not an issue, since it allows for support regeneration, unlike immobilization by covalent binding.

Ion-exchanger resins are typical supports for ionic binding; among them are derivatives of cross-linked polysaccharides, namely, carboxymethyl- (CM-) cellulose, CM-Sepharose, diethylaminoethyl- (DEAE-) cellulose, DEAE-Sephadex, quaternary aminoethyl anion exchange- (QAE-) cellulose, QAE-dextran, QAE-Sephadex; derivatives of synthetic polymers, namely, Amberlite, Diaion, Dowex, Duolite; and resins coated with ionic polymers, namely, polyethylenimine (PEI).

Recent examples include the immobilization of invertase in Dowex, in Duolite, in poly(glycidyl methacrylate-co-methyl methacrylate beads grafted with PEI, and in epoxy(amino) Sepabeads; lactase immobilization in PEI-grafted Sepabeads; fructosyltransferase in DEAE-cellulose for the production of fructosyl disaccharides; glucose isomerase in DEAE-cellulose or in Indion 48-R; glucoamylase onto SBA-15 silica and in epoxy(amino) Sepabeads. Ionic binding to Sepabeads-like supports has acknowledged multipoint attachment nature. Enzyme molecules can be modified chemically or genetically modified to enhance immobilization efficiency, an approach followed by Kweon and coworkers, who obtained a cyclodextrin glycosyltransferase fused with 10 lysine residues to improve ionic binding to SP-Sepharose.

Carrier-free macroparticles, where a bifunctional reagent (namely, glutaraldehyde), is used to cross-link enzyme aggregates (CLEAs) or crystals (CLECs), leading to a biocatalyst displaying highly concentrated enzyme activity, high stability and low production costs.

The use of CLEAs is favoured given the lower complexity of the process. This approach is recent, as compared with entrapment and binding to a solid carrier, and there are still relatively few examples of its application to enzymes used in the area of food processing. Among those are following.

First is the immobilization of Pectinex Ultra SP-L, a commercial enzyme preparation containing pectinase, xylanase, and cellulose activities.

The CLEA biocatalyst displayed a slight (30%) in the Vmax, maximal reaction rate/KM ratio, but a significant enhancement in thermal stability (a roughly 10-fold increase in half-life), when the pectinase activity of the immobilized biocatalyst was compared with the free form.

Second is the immobilization of lactase for the hydrolysis of lactose, where, under similar operational conditions as for the free enzyme, the CLEA yielded 78% monosaccharides in 12 h as compared to 3.9% of the free form. Third, CLEAs of glucoamylase, formed by either glutaraldehyde or diimidates, namely, dimethylmalonimidate, dimethylsuccinimidate, and dimethylglutarimidate, led to biocatalysts with improved thermal stability as compared to the free form (over 2-fold increase in half-lives).

Fourth, CLEAs of wild type and two mutant levansucrases were assayed for oligosaccharides/levan and for fructosyl-xyloside synthesis. Although the specific activity of the three free enzymes was 1.25- to 3-fold higher than the corresponding CLEAs, these displayed a 40-

to 200-fold higher specific activity than the equivalent Eupergit-C-immobilized enzyme preparations.

Furthermore, all CLEA preparations displayed enhanced thermal stability when compared with the corresponding free enzymes.

Fifth are CLECs of glucose isomerase, aimed at the conversion of glucose into fructose for the production of high fructose corn syrup. When placed in a packed-bed, the resulting enzyme preparation allowed for flow rates that matched or even exceeded those processed by commercially available enzyme preparations (either free, carrier free, or carrier-bound), while achieving the same 45% yield in fructose, under similar operational condition.

Sixth, CLECs of glucose isomerase packed in a column were also used for the concentration/purification of xylitol from dilute or impure solutions. The approach was based on the high specificity of the enzyme crystals towards xylitol, allowing its separation from other sugars, including the natural substrates, xylose and glucose.

Recovery of the adsorbed xylitol was achieved by elution with $CaCl_2$ solutions, with Ca^{2+} being acknowledged to inactivate glucose isomerase.

Each method for enzyme immobilization has a unique nature. Therefore, despite the potential of immobilization to improve enzyme performance by enhancing activity, stability, or specificity, no specific approach tackles simultaneously these different features.

A careful evaluation and characterization of the methodology addressed is thus required, which can be significantly fastened by high-throughput approaches. Again, the feasibility of its application to reactor configuration and mode of operation has also to be considered in the selection process of the most adequate immobilized biocatalyst for a given bioconversion.

Typical Bioreactors

The most common form of enzymatic reactors for continuous operation is the packed-bed setup, basically a cylindrical column holding a fixed bed of catalyst particles.

These should not have sizes below 0.05mm, in order to keep the pressure drop within reasonable limits. Commercially available carriers such as Eupergit C have particle sizes of roughly 0.1 mm. Commonly operated in down-flow mode, the range of flow rates used must be such as to provide a compromise between reasonable pressure drop, minimal diffusion layer and high conversion yield. Minimization of external mass-transfer resistances with enhanced flow rates can be considered, leading to the fluidized-bed reactor.

This is basically a variation of the packed-bed reactor, but operated in up-flow mode, where the biocatalyst particles are not in close contact which each other; hence, pressure drop is low, and accordingly are pumping costs.

The residence time allowed by the flow rates required for fluidization may however result in low conversion yields. This can be overcome by operating a battery of reactor or by operation in recycle mode. Bioconversions with free enzymes are carried out in stirred tanks.

When on their own, they are restricted to batch mode, but when coupled to a membrane setup with suitable cutoff, they can be integrated in a continuous process, since the enzymes are rejected by the membrane, which acts as an immobilization device, whereas the product (and unconverted substrate) freely permeates. Shear stress induced by stirring creates a hazardous environment for immobilized biocatalysts, particularly when hydrogels are considered, since they are prone to abrasion. In order to overcome this, a basket reactor was developed, but is seldom used, possibly due to mass transfer resistances associated.

Conclusions and Future Perspectives

The integration of enzymes in food and feed processes is a well-established approach, but evidence clearly shows that dedicated research efforts are consistently being made as to make this application of biological agents more effective and/or diversified. These endeavors have been anchoring in innovative approaches for the design of new/improved biocatalysts, more stable (to temperature and pH), less dependent on metal ions and less susceptible to inhibitory agents and to aggressive environmental conditions, while maintaining the targeted activity or evolving novel activities.

This is of particular relevance for application in the food and feed sector, for it allows enhanced performance under operational conditions that minimize the risk of microbial contamination. It also favours process integration, by allowing the concerted use of enzymes that naturally have diverse requirements for effective application. Such progresses have been made through the ever-continuing developments in molecular biology, the accumulated evolutionary enzyme engineering expertise, the (bio) computational tools, and the implementation of high-throughput methodologies, with high level of parallelization, enabling the efficient and timely screening/characterization of the biocatalysts. Alongside with these strategies, the immobilization of enzymes has also been a key supporting tool for rendering these proteins fit for industrial application, while simultaneously enabling the improvement of their catalytic features. Again, and despite the developments made in this particular field, there is still the lack of a set of unanimously applicable rules for the selection of carrier and method of enzyme immobilization, which furthermore encompass both technical and economic requirements. The latter can be particularly restrictive in the food and feed sector, since most products are of relatively low added value.

Therefore, there is no universal support and method for enzyme immobilization aimed at application in food and feed (let alone the overall range of possible fields of use), and the immobilized biocatalyst fit for a given process and product may be totally unsuitable for another. Given the diversity of enzyme nature and applications this pattern is unlikely to be reversed. Hence, it can be foreseen that efforts will be towards the development of immobilized biocatalyst with suitable chemical, physical, and geometric characteristics, which can be produced under mild condition, that can be used in different reactor configurations and that comply with the economic requirements for large-scale application. All these strategies either isolated or preferably suitably integrated have been put into practice in food and feed, to improve existing processes or to implement new ones, with the latter often combined with the output of new goods, resulting from novel enzymatic activities. Given the recent developments in this field, this trend is foreseen to be further implemented.

Chapter 3

Application of Modern Biotechnology in Food Processing

Contrary to its name, biotechnology is not a single technology. Rather it is a group of technologies that share two (common) characteristics — working with living cells and their molecules and having a wide range of practice uses that can improve our lives.

Biotechnology can be broadly defined as "using organisms or their products for commercial purposes." As such, (traditional) biotechnology has been practices since he beginning of records history. (It has been used to:) bake bread, brew alcoholic beverages, and breed food crops or domestic animals. But recent developments in molecular biology have given biotechnology new meaning, new prominence, and new potential. It is (modern) biotechnology that has captured the attention of the public. Modern biotechnology can have a dramatic effect on the world economy and society.

One example of modern biotechnology is genetic engineering. Genetic engineering is the process of transferring individual genes between organisms or modifying the genes in an organism to remove or add a desired trait or characteristic. Examples of genetic engineering are described later in this document. Through genetic engineering, genetically modified crops or organisms are formed. These GM crops or GMOs are used to produce biotech-derived foods. It is this specific type of modern biotechnology, genetic engineering, that seems to generate the most attention and concern by consumers and consumer groups. What is interesting is that modern biotechnology is far more precise than traditional forms of biotechnology and so is viewed by some as being far safer.)

How does Modern Biotechnology Work?

All organisms are made up of cells that are programmed by the same basic genetic material, called DNA (deoxyribonucleic acid). Each unit of DNA is made up of a combination of the following nucleotides — adenine (A), guanine (G), thymine (T), and cytosine (D) — as well as a sugar and a phosphate. These nucleotides pair up into strands that twist together into a spiral structure call a "double helix." This double helix is DNA. Segments of the DNA tell individual cells how to produce specific proteins. These segments are genes. It is the presence or absence of the specific protein that gives an organism a trait or characteristic. More than 10,000 different genes are found in most plant and animal species. This total set of genes for an organism is organized into chromosomes within the cell nucleus. The process by which a multi-cellular organism develops from a single cell through an embryo stage into an adult is ultimately controlled by the genetic information of the cell, as well as interaction of genes and gene products with environmental factors.

When cells reproduce, the DNA strands of the double helix separate. Because nucleotide A always pairs with T and G always pairs with C, each DNA strand serves as a precise blueprint for a specific protein. Except for mutations or mistakes in the replication process, a single cell is equipped with the information to replicate into millions of identical cells. Because all organisms are made up of the same type of genetic material (nucleotides A, T, G, and C), biotechnologists use enzymes to cut and remove DNA segments from one organism and recombine it with DNA in another organism. This is called recombinant DNA (rDNA) technology, and it is one of the basic tools of modern biotechnology. rDNA technology is the laboratory manipulation of DNA in which DNA, or fragments of DNA from different sources, are cut and recombined using enzymes. This recombinant DNA is then inserted into a living organism. rDNA technology is usually used synonymously with genetic engineering. rDNA technology allows researchers to move genetic information between unrelated organisms to produce desired products or characteristics or to eliminate undesirable characteristics.

Genetic engineering is the technique of removing, modifying or adding genes to a DNA molecule in order to change the information it contains. By changing this information, genetic engineering changes the type or amount of proteins an organism is capable of producing. Genetic engineering is used in the production of drugs, human gene therapy, and the development of improved plants. For example, an "insect protection" gene (Bt) has been inserted into several crops - corn, cotton, and potatoes - to give farmers new tools for integrated pest management. Bt corn is resistant to European corn borer. This inherent resistance thus reduces a farmers pesticide use for controlling European corn borer, and in turn requires less chemicals and potentially provides higher yielding Agricultural Biotechnology.

Although major genetic improvements have been made in crops, progress in conventional breeding programs has been slow. In fact, most crops grown in the US produce less than their full genetic potential. These shortfalls in yield are due to the inability of crops to tolerate or adapt to environmental stresses, pests, and diseases. For example, some of the world's highest yields of potatoes are in Idaho under irrigation, but in 1993 both quality and yield were severely reduced because of cold, wet weather and widespread frost damage during June. Some of the world's best bread wheats and malting barleys are produced in the north-central states, but in 1993 the disease Fusarium caused an estimated $1 billion in damage. Scientists have the ability to insert genes that give biological defense against diseases and insects, thus reducing the need for chemical pesticides, and they will soon be able to convey genetic traits that enable crops to better withstand harsh conditions, such as drought. The International Laboratory for Tropical Agricultural Biotechnology (ILTAB) is developing transformation techniques and applications for control of diseases caused by plant viruses in tropical plants such as rice, cassava and tomato. In 1995, ILTAB reported the first transfer through biotechnology of a resistance gene from a wild species of rice to a susceptible cultivated rice variety. The transferred gene expressed resistance to Xanthomonas oryzae, a bacterium which can destroy the crop through disease. The resistant gene was transferred into susceptible rice varieties that are cultivated on more than 24 million hectares around the world.

Benefits can also be seen in the environment, where insect-protected biotech crops reduce the need for chemical pesticide use. Insect-protected crops allow for less potential exposure of farmers and ground-water to chemical residues, while providing farmers with season-long control. Also by reducing the need for pest control, impacts and resources spent on the land are less, thereby preserving the topsoil.

Major advances also have been made through conventional breeding and selection of livestock, but significant gains can still be made by using biotechnology. Currently, farmers in the U.S spend $17 billion dollars on animal health. Diseases such as hog cholera and pests such as screwworm have been eradicated. Uses of biotechnology in animal production include development of vaccines to protect animals from disease, production of several calves from one embryo (cloning), increase of animal growth rate, and rapid disease detection.

Modern biotechnology has offered opportunities to produce more nutritious and better tasting foods, higher crop yields and plants that are naturally protected from disease and insects. Modern biotechnology allows for the transfer of only one or a few desirable genes, thereby permitting scientists to develop crops with specific beneficial traits and reduce undesirable traits.

Traditional biotechnology such as cross-pollination in corn produces numerous, non-selective changes. Genetic modifications have produced fruits that can ripen on the vine for better taste, yet have longer shelf lives through delayed pectin degradation. Tomatoes and other produce containing increased levels of certain nutrients, such as vitamin C, vitamin E, and or beta carotene, and help protect against the risk of chronic diseases, such as some cancers and heart disease. Similarly introducing genes that increase available iron levels in rice three-fold is a potential remedy for iron deficiency, a condition that effects more than two billion people and causes anemia in about half that number. Most of the today's hard cheese products are made with a biotech enzyme called chymosin. This is produced by genetically engineered bacteria which is considered more purer and plentiful than it's naturally occurring counterpart, rennet, which is derived from calf stomach tissue.

In 1992, Monsanto Company successfully inserted a gene from a bacterium into the Russet Burbank potato. This gene increases the starch content of the potato. Higher starch content reduces oil absorption during frying, thereby lowering the cost of processing french fries and chips and reducing the fat content in the finished product. This product is still awaiting final development and approval.

Modern biotechnology offers effective techniques to address food safety concerns. Biotechnical methods may be used to decrease the time necessary to detect foodborne pathogens, toxins, and chemical contaminants, as well as to increase detection sensitivity. Enzymes, antibodies, and microorganisms produced using rDNA techniques are being used to monitor food production and processing systems for quality control.

Biotechnology can compress the time frame required to translate fundamental discoveries into applications. This is done by controlling which genes are altered in an organized fashion. For example, a known gene sequence from a corn plant can be altered to improve yield, increase drought tolerance, and produce insect resistance (Bt) in one generation. Conventional breeding techniques would take several years. Conventional breeding techniques would require that a field of corn is grown and each trait is selected from individual stalks of corn. The ears of corn from selected stalks with each desired trait (e.g, drought tolerance and yield performance) would then be grown and combined (cross-pollinated). Their offspring (hybrid) would be further selected for the desired result (a high performing corn with drought tolerance). With improved technology and knowledge about agricultural organisms, processes, and ecosystems, opportunities will emerge to produce new and improved agricultural products in an environmentally sound manner.

In summary, modern biotechnology offers opportunities to improve product quality, nutritional content, and economic benefits. The genetic make-up of plants and animals can be modified by either insertion of new useful genes or removal of unwanted ones. Biotechnology is changing the way plants and animals are grown, boosting their value to growers, processors, and consumers.

Industrial Biotechnology

Industrial biotechnology applies the techniques of modern molecular biology to improve the efficiency and reduce the environmental impacts of industrial processes like textile, paper and pulp, and chemical manufacturing. For example, industrial biotechnology companies develop biocatalysts, such as enzymes, to synthesize chemicals. Enzymes are proteins produced by all organisms. Using biotechnology, the desired enzyme can be manufactured in commercial quantities.

Commodity chemicals (e.g., polymer-grade acrylamide) and speciality chemicals can be produced using biotech applications. Traditional chemical synthesis involves large amounts of energy and often-undesirable products, such as HCl. Using biocatalysts, the same chemicals can be produced more economically and more environmentally friendly. An example would be the substitution of protease in detergents for other cleaning compounds. Detergent proteases, which remove protein impurities, are essential components of modern detergents.

They are used to break down protein, starch, and fatty acids present on items being washed. Protease production results in a biomass that in turn yields a useful by-product- an organic fertilizer. Biotechnology is also used in the textile industry for the finishing of fabrics and garments. Biotechnology also produces biotech-derived cotton that is warmer, stronger, has improved dye uptake and retention, enhanced absorbency, and wrinkle- and shrink-resistance. Some agricultural crops, such as corn, can be used in place of petroleum to produce chemicals.

The crop's sugar can be fermented to acid, which can be then used as an intermediate to produce other chemical feedstocks for various products. It has been projected that 30% of the world's chemical and fuel needs could be supplied by such renewable resources in the first half of the next century. It has been demonstrated, at test scale, that biopulping reduces the electrical energy required for wood pulping process by 30%.

Environmental Biotechnology

Environmental biotechnology is the used in waste treatment and pollution prevention. Environmental biotechnology can more efficiently clean up many wastes than conventional methods and greatly reduce our dependence on methods for land-based disposal.

Every organism ingests nutrients to live and produces by-products as a result. Different organisms need different types of nutrients. Some bacteria thrive on the chemical components of waste products. Environmental engineers use bioremediation, the broadest application of environmental biotechnology, in two basic ways. They introduce nutrients to stimulate the activity of bacteria already present in the soil at a waste site, or add new bacteria to the soil. The bacteria digest the waste at the site and turn it into harmless by-products. After the bacteria consume the waste materials, they die off or return to their normal population levels in the environment.

Bioremediation, is an area of increasing interest. Through application of biotechnical methods, enzyme bioreactors are being developed that will pretreat some industrial waste and food waste components and allow their removal through the sewage system rather than through solid waste disposal mechanisms. Waste can also be converted to biofuel to run generators. Microbes can be induced to produce enzymes needed to convert plant and vegetable materials into building blocks for biodegradable plastics.

In some cases, the by-products of the pollution-fighting microorganisms are themselves useful. For example, methane can be derived from a form of bacteria that degrades sulphur liquor, a waste product of paper manufacturing. This methane can then be used as a fuel or in other industrial processes.

Human Applications

Biotechnical methods are now used to produce many proteins for pharmaceutical and other specialized purposes. A harmless strain of Escherichia coli bacteria, given a copy of the gene for human insulin, can make insulin. As these genetically modified (GM) bacterial cells age, they produce human insulin, which can be purified and used to treat diabetes in humans. Microorganisms can also be modified to produce digestive enzymes. In the future, these microorganisms could be colonized in the intestinal tract of persons with digestive enzyme insufficiencies. Products of modern biotechnology include artificial blood vessels from collagen tubes coated with a layer of the anticoagulant heparin.

Gene therapy – altering DNA within cells in an organism to treat or cure a disease – is one of the most promising areas of biotechnology research. New genetic therapies are being developed to treat diseases such as cystic fibrosis, AIDS and cancer. DNA fingerprinting is the process of cross matching two strands of DNA. In criminal investigations, DNA from samples of hair, bodily fluids or skin at a crime scene are compared with those obtained from the suspects. In practice, it has become one of the most powerful and widely known applications of biotechnology today. Another process, polymerase chain reaction (PCR), is also being used to more quickly and accurately identify the presence of infections such as AIDS, Lyme disease and Chlamydia.

Paternity determination is possible because a child's DNA pattern is inherited, half from the mother and half from the father. To establish paternity, DNA fingerprints of the mother, child and the alleged father are compared. The matching sequences of the mother and the child are eliminated from the child's DNA fingerprint; what remains comes from the biological father. These segments are then compared for a match with the DNA fingerprint of the alleged father.

DNA testing is also used on human fossils to determine how closely related fossil samples are from different geographic locations and geologic areas. The results shed light on the history of human evolution and the manner in which human ancestors settled different parts of the world.

Biotechnology for the 21st Century

Experts in United States anticipate the world's population in 2050 to be approximately 8.7 billion persons. The world's population is growing, but its surface area is not. Compounding the effects of population growth is the fact that most of the earth's ideal farming land is already being utilized. To avoid damaging environmentally sensitive areas, such as rain forests, we need to increase crop yields for land currently in use. By increasing crop yields, through the use of biotechnology the constant need to clear more land for growing food is reduced.

Countries in Asia, Africa, and elsewhere are grappling with how to continue feeding a growing population. They are also trying to benefit more from their existing resources. Biotechnology holds the key to increasing the yield of staple crops by allowing farmers to reap bigger harvests from currently cultivated land, while preserving the land's ability to support continued farming.

Malnutrition in underdeveloped countries is also being combated with biotechnology. The Rockefeller Foundation is sponsoring research on "golden rice", a crop designed to improve nutrition in the developing world. Rice breeders are using biotechnology to build Vitamin A into the rice. Vitamin A deficiency is a common problem in poor countries. A second phase of the project will increase the iron content in rice to combat anemia, which is widespread problem among women and children in underdeveloped countries. Golden rice, expected to be for sale in Asia in less than five years, will offer dramatic improvements in nutrition and health for millions of people, with little additional costs to consumers.

Similar initiatives using genetic manipulation are aimed at making crops more productive by reducing their dependence on pesticides, fertilizers and irrigation, or by increasing their resistance to plant diseases.

Increased crop yield, greater flexibility in growing environments, less use of chemical pesticides and improved nutritional content make agricultural biotechnology, quite literally, the future of the world's food supply.

Concerns about Biotechnology

As biotechnology has become widely used, questions and concerns have also been raised. The most vocal opposition has come from European countries. One of the main areas of concern is the safety of genetically engineered food.

In assessing the benefits and risks involved in the use of modern biotechnology, there are a series of issues to be addressed so that informed decisions can be made. In making value judgments about risks and benefits in the use of biotechnology, it is important to distinguish between technology-inherent risks and technology-transcending risks. The former includes assessing any risks associated with food safety and the behaviour of a biotechnology-based product in the environment. The latter involve the political and social context in which the technology is used, including how these uses may benefit or harm the interests of different groups in society.

The health effects of foods grown from genetically engineered crop depend on the composition of the food itself. Any new product may have either beneficial or occasional harmful effects on human health. For example, a biotech-derived food with a higher content of digestible iron is likely to have a positive effect if consumed by iron-deficient individuals. Alternatively, the transfer of genes from one species to another may also transfer the risk for exposure to allergens.

These risks are systematically evaluated by FDA and identified prior to commercialization.

Individuals allergic to certain nuts, for example, need to know if genes conveying this trait are transferred to other foods such as soybeans. Labelling would be required if such crops were available to consumers.

Among the potential ecological risks identified are increased weediness, due to cross- pollination from genetically modified crops spreads to other plants in nearby fields. This may allow the spread of traits such as herbicide-resistance to non-target plants that could potentially develop into weeds. This ecological risk is assessed when deciding if a plant with a given trait should be released into a particular environment, and if so, under what conditions.

Other potential ecological risks stem from the use of genetically modified corn and cotton with insecticidal genes from Bacillus thuringiensis (Bt genes). This may lead to the development of resistance to Bt in insect populations exposed to the biotech-derived crop. There also may be risks to non-target species, such as birds and butterflies, from the plants with Bt genes. The monitoring of these effects of new crops in the environment and implementation of effective risk management approaches is an essential component of further research. It is also important to keep all risks in perspective by comparing the products of biotechnology and conventional agriculture.

The reduction of biodiversity would represent a technology-transcending risk. Reduced biological diversity due to destruction of tropical forests, conversion of land to agriculture, overfishing, and the other practices to feed a growing world population is a significant loss far more than any potential loss of biodiversity due to biotech-derived crop varieties. Improved governance and international support are necessary to limit loss of biodiversity.

What we know from our understanding of science and more than a decade of experience with biotech-derived plants is the following: There is no evidence that genetic transfers between unrelated organisms pose human health concerns that are different from those encountered with any new plant or animal variety. The risks associated with biotechnology are the same as those associated with plants and microbes developed by conventional methods.

Consumer and Food Industry Perspectives

Survey research over the past decade shows that biotechnology is not likely to become an important issue for most American consumers. Consumers find biotechnology acceptable when they believe it offers benefits and it is safe. Surveys have consistently found that a majority of American consumers are willing to buy insect-protected food crops developed through biotechnology that use fewer chemical pesticides, as well as more nutritious foods. American consumers also appreciate the role that biotechnology can play in feeding the world. Research shows that European consumers are much less supportive of all biotechnology applications.

Surveys since 1992 show that relatively few U.S. consumers have heard or read much about biotechnology. News about the cloned sheep pushed awareness to 50 percent in March 1997. Surveys in the first three months of 2000 show that awareness has fallen back to just over one-third in the United States. Such trends reflect the fact that most people get their information about biotechnology from the media. Unfortunately, many consumers also do not understand some fundamental principles of biology. European consumer awareness is somewhat higher, but knowledge is still low.

Media coverage in the United States has generally been balanced (which helps account for our relatively high levels of acceptance). This is in sharp contrast to the European media, which have played upon fear of the unknown. The European media have also tended to accept opponents' claims without question. Another issue is that many people no longer have a connection to agriculture. In fact, research has shown that many consumers are unaware that all foods are derived from plants or animals that already have been genetically modified through traditional (but imprecise) breeding methods. American consumers look to health professionals and scientific experts for credible information, but place relatively little trust in the activists who oppose biotechnology. Research shows that acceptance increases significantly when American consumers learn that organizations such as the National Academy of Sciences and the U.S. Food and Drug Administration have determined that biotech-derived foods are safe. In contrast, European consumers express the most trust in those groups that oppose biotechnology. They have much less confidence in government, industry, or even scientists. American culture is more supportive and rewarding of new technology.

Europeans tend to view food differently from U.S. consumers. In fact, some Europeans reject all American food products. Europeans also want to protect their small farms to maintain open space and rural employment. Such forces underlie much of the European anxiety about agricultural biotech - especially since it is seen as an "American invention."

Most of the industry leaders interviewed are quite enthusiastic about the benefits of biotechnology — especially in terms of increased food availability, enhanced nutrition, and environmental protection. Most feel that biotechnology has already provided benefits to consumers. Almost all recognize that foods developed through biotechnology have already been part of consumers' everyday diet. They clearly do not agree with most of the opponents' claims and tend to have almost no trust in such groups. Their main concerns involve lack of consumer acceptance — not the safety of the foods. They express high levels of confidence in the science and the regulatory process. In fact, almost none feel that biotechnology should not be used because of uncertain, potential risks. Most food industry leaders do not feel it is necessary to have special labels on biotech-derived foods. They express concerns that such labels would be perceived as a warning by consumers. They also worried that the need to segregate commodities would pose financial and logistical burdens on everyone in the system - including consumers. Food industry leaders recognize a major need to educate the public about biotechnology. They look to third parties, such as university and government scientists to provide such leadership.

Research shows that consumers will accept biotech foods if they see a benefit to themselves or society and if the price is right. Their responses to foods developed through biotechnology are basically the same as for any other food - taste, nutrition, price, safety and convenience are the major factors that influence our decisions about which foods to eat. How seeds and food ingredients are developed will only be relevant for a relatively small group of concerned, consumers. The potential for public concerns has led several food companies to change their products to avoid biotech-derived ingredients. For example, Gerber Foods received threats from Green peace because they had determined the company was using biotech-derived food ingredients (mainly soy). The company firmly believes that the biotech foods are safe to consume. Gerber agreed to drop some of its existing corn and soybean suppliers in favor of ones that can produce crops that are not genetically altered. It became an issue that is suddenly confronting other food companies.

A private manufacturer in California, called Healthy Times Natural Food has switched from Canola oil (which sometimes is genetically modified) to safflower oil after facing questions from Green peace. The controversy is due in part to the fact that the organic industry is using public concern as a tool for marketing their products as free of biotech ingredients.

National and International Biotechnology Policy

National governments and international policy making bodies rely on food scientists and others to develop innovations that will create marketable food products and increase food supplies.

Governments also rely on scientific research because they are responsible for setting health and safety standards regarding new developments. International organizations can suggest policy approaches and help develop international treaties that are ratified by national governments.

Economic success in the competitive international market demands that food production become more efficient and profitable. National governments and international organizations support food biotechnology as a means to avoid global food shortages. Many policy making bodies are also trying to balance support of the food biotechnology industry with public calls for their regulation. Such regulations are necessary to protect public health and safety, to promote international trade, conserve natural resources, and account for ethical issues.

The majority of processed foods on the market contain soy or corn ingredients that come from GM plants. To date none have posed a food safety risk. The chief safety concerns are the potential to alter nutrient content or introduce allergens. Federal agencies involved in biotechnology regulation include the U.S. Department of Agriculture (USDA) which evaluates agricultural production processes for all foods; the Food and Drug Administration (FDA), which evaluates whole non-animal foods (seafood), food ingredients, and food additives; and the Environmental Protection Agency (EPA), which evaluates plants with insecticidal properties.

Developers of GM plants and biotech-derived foods are required to consult with FDA prior to the commercialization of the product. This consultation procedure entails a science-based safety assessment of the product that focuses on the protection of the consumer, developer, and the environment. Thus developers, have a strong incentive to cooperate fully with FDA and the other agencies prior to marketing their products.

Labelling Issues

Labelling food derived from GM plants and animals is an important, but complex issue. Some consumers and consumer groups believe they have a right to know whether biotechnology was used to produce food. Others believe labelling is not necessary if foods are essentially equivalent in composition. Food labels are regulated by FDA and, in some cases, by USDA. Regulatory agencies are concerned with ensuring that food labels are both true and not misleading. In accordance with their statutory mandate, the FDA has determined that a food product should be labelled as a product of biotechnology only if it has been changed in some significant way. The U.S. Court of Appeals recently reaffirmed that the FDA's labelling policy is correct. Based on the input from three public hearings, the FDA has recognized that the most effective way to allow for informed choice will be a system of voluntary labelling for foods not produced through biotechnology. If the demand for "GM-free" food is real, a market will become viable.

In this case, meaningful choice can be provided to concerned consumers without imposing costs on or denying benefits to the majority of consumers, who support biotechnology. This may be the only viable and cost-effective option for the food industry (given the costs and difficulties associated with segregation and identity preservation.)

National surveys conducted with American consumers between 1997 and 2000 found about three-quarters of consumers supported this FDA labelling policy. There is evidence from focus groups that US consumers are already overwhelmed by the level of detail on food labels and mainly look for relevant information about nutrition. Recent focus groups have demonstrated that the wording on labels has a significant effect on consumer understanding and acceptance of biotechnology. Any label messages need to be simple and clear. However, labelling will be counterproductive without education.

It is important to understand when FDA will and won't require a product label. The same labelling laws that apply to all other foods and food ingredients apply to products of food biotechnology. Requirements for all food labels mandate proper identification of products and notice of health or safety concerns. Labelling would be required in the following instances – but not simply because the products were made using biotechnology.

* Potential food allergy is an example of a health or safety risk that would mandate a product label (e.g., if a gene has been transferred from a known allergen, such as a peanut).
* When substantial changes have been made to the expected nutritional composition of a food (e.g., vitamins have been added to a particular type of produce).
* If the common identity of a food product has been changed (e.g., if the composition of a fruit or vegetable becomes significantly different it would have to have a different name).

Labelling of biotech-derived foods will be required in some countries (such as Japan and the European Union) for cultural and religious reasons. These countries also believe labelling should be required simply because the consumers want to know more about the content of the food or how it was produced. Some have suggested that such labelling initiatives have been driven by economic and political interests rather than by scientific need. Consumers in these countries will ultimately end up paying even higher prices for their food as a result of such strict labelling policies.

The applications of biotechnology are so broad, and the advantages so compelling, that virtually every industry is using this technology. Developments are underway in areas as diverse as pharmaceuticals, diagnostics, textiles, aquaculture, forestry, chemicals, household products, environmental cleanup, food processing and forensics to name a few. Biotechnology is enabling these industries to make new or better products, often with greater speed, efficiency and flexibility. Biotechnology holds significant promise to the future but certain amount of risk is associated with any area. Biotechnology must continue to be carefully regulated so that the maximum benefits are received with the least risk.

Biotechnology is at a crossroads in terms of public acceptance. Many Americans have not yet formed a solid opinion on this complex

Application of Modern Biotechnology in Food Processing

issue. International developments over the next few years will certainly have a major influence on the long-term viability of biotechnology. The future of the world food supply depends upon how well scientists, government, and the food industry are able to communicate with consumers about the benefits and safety of the technology.

Several major initiatives are under way to strengthen the regulatory process and to communicate more effectively with consumers. Both the USDA and FDA have opened their regulatory systems to outside review and public comment. The biotechnology industry, university scientists and others are also conducting educational programs.

These should further strengthen consumer confidence. This partnership among the public and private sectors will support these emerging technologies that will prove vital to the U.S. economy and the developing world in the new millennium. Even Europe will soon find the real benefits of biotechnology compelling.

The 6 Top Trends in Food Processing

When it comes to food and beverage products, one rule of thumb defines a true trend: Real trends don't come and go; they grow – over years, decades or even longer. For processors, fads can still bring in big money, but the risk is high. Ask any food executive caught last year with a million bucks' worth of low-carb inventory.

In between fads and trends are trendlets. These are the bubbles that pop up within a trend and are worth noting because they can provide a hook for food and beverage processors to hang their R&D hats on.

We went shopping for real trends. We sifted through numerous media reports on what was hot, comments from the editorial advisory boards of both *Food Processing* and our sister magazine *Wellness Foods* and the wisdom shared at the many food shows we attend each year. We culled a huge list down to six essential trends.

New Food Products Resource Center

Food Processing reports on a variety of new food products. Find consumer market research for particular products, monthly product roll-out roundups and other monthly features in one place.

We've pegged Organic as the biggest and most significant trend for processors to watch. No great surprise, but more important than our No. 2 pick, Health & Wellness? Yes. Now entering its second generation of double-digit growth, organic is on another big upswing.

In simple terms, although the Health & Wellness category has a larger footprint than Organic – and the two often are lumped together erroneously – the fact is, any food product can be formulated to sport an organic label. However, not all products can wear the "healthy" tag.

Example: A rapidly growing niche in the beverage world is that of organic beer, wine and spirits. Discounting the studies showing possible benefits of moderate alcohol intake, organic vodka won't likely get a health claim.

Health & Wellness is the second and broadest of the hot trends. It encompasses such huge components as diabetes and obesity, kids' health, food safety, women's health, allergies and immunity as well as the fringe issues of "well-being" and "energy."

As a trend, Age Awareness certainly overlaps with Health & Wellness, especially as the latter concerns our aging population. But there are numerous non-health aspects for processors to consider as they help our 77 million baby boomers segue into their dotage.

Portion Control is our No. 4 trend. In some ways, it's just a health tool. But this year, especially, it deserves its own category because it constitutes a merging of health with the perennial trend for more convenience. And it's one trend numerous processors from all categories are jumping on.

Globalization is No. 5. Immigration controversies aside, Asian, Hispanic, African American and other ethnic minorities will make up more than 35 percent of the U.S. population in about five years, according to estimates derived from U.S. Census Bureau data. About half that figure will be Hispanics. But globalization and ethnic influences are more than population figures. Today's businesses are international more often than not. The cultural traffic and instant global information (via mass media and the Internet) mean rapid diffusion of once regional preferences.

Rounding out the six major processing trends are Kosher and halal certification. The strange bedfellows continue to be prolific growth areas, with Kosher still progressing at double-digit rates and halal experiencing a sudden and major growth spurt.

Organic = Healthy

That equation is not necessarily true, but the message is so ingrained in the minds of millions of consumers that the math cannot be ignored.

U.S. organic food sales totaled nearly $14 billion in 2005, according to the Organic Trade Assn, Greenfield, Mass. (OTA). Although this represents a mere 2.5 percent of all retail food sales, that total is a 31 percent increase over 2003 figures. According to OTA, sales of organic foods are expected to reach nearly $16 billion by the end of 2006.

"These findings show there is continued strong growth for organic products," says Caren Wilcox, OTA's executive director. OTA statistics show organic food categories experiencing the greatest growth during 2005 included meat (55.4 percent), condiments (24.2 percent) and dairy products (23.5 percent). Even Fido and Garfield are going organic: One of the fastest-growing organic categories during 2005 was pet food (46 percent).

"We're at a point where demand for organic product exceeds supply," says Robert Schueller, director of public relations for Melissa's/World Variety Produce Inc., Los Angeles. "Organic everything is hot, and it's not just produce. Most of the top retailers in the country offer organic products in their stores, and many foodservice establishments have taken notice, offering organic foods and ingredients more often on the menu."

One of the largest distributors of variety organic produce in the country, Melissa's offers more than 350 organic produce items. Gearing up for a larger-than-normal increase in demand, Schueller notes the number will reach about 400 SKUs by year's end. Since the 2004 start of its Good Life Food brand of organic processed items, sales have grown an average of 20 percent a year. Forecasts double that figure for 2006.

It's not just the patchouli set that sees the advantages to organic foods. Surveys show the majority of Americans are concerned about what's in their foods, where those foods come from and potential health risks from pesticides and chemicals in the food chain.

David Johnson, president-North America commercial at Kraft Foods Inc., Northfield, Ill, has been quoted describing the organic food trend as "a freight train that's going to pick up steam." As the second largest food company in North America, the company is in the position to help make this a self-fulfilling prophecy.

According to a recent study by ACNielsen, Chicago, organic products topped the list of "best performing" items in the "good-for-you" product segments. But organic is its own trend, extending beyond foods, beverages and pet foods. The category has acquired such trendlets as environmental consciousness and sustainability, Fair Trade, local production, energy conservation and "natural," minimally processed or stripped-down formulations. Although not making an organic claim, Cadbury Schweppes PLC, Plano, Tex., recently reformulated and remarketed its 7Up beverage as a five-ingredient, "natural" product.

Get Well Soon

The twin epidemics of obesity and diabetes dominate the health and wellness category. No day passes without the mention of one, the other or both on television, radio or in newspapers. But in general, between one-fourth and one-third of consumers make food choices based on health for some reason.

This is a trend that plays directly to our desire to ingest specific foods or beverages for the purpose of preventing or palliating a disease or condition. The reason antioxidants, botanical extracts and the whole foods (berries, teas, soy) that contain them are critical underpinnings of this trend is because of the promise such items hold to improve how you feel and perform.

From an ingredient standpoint, health and wellness concerns offer the best variety of options for processors. A manufacturer developing a product in this market has literally thousands of botanical extracts, antioxidants, phytochemicals, carbohydrate compounds (such as sugars, starches and fibers), protein compounds or fractions and healthy oils from which to choose. The trend to address health reached its mainstream "tipping point" in food processing with the decision in 2004 by America's second largest cereal maker, General Mills Inc., to reformulate all of its breakfast cereals to be based on whole grains. There certainly were defining moments along the way – fortification of flour and breads with folate, calcium enrichment of juice and other "health-value added" movements all are good examples. But the Minneapolis-based company was the first processing giant to change its entire line of products in such a manner.

Meanwhile, emphasis is shifting away from dieting and related fat and calorie-count issues. In a survey by Mintel International, Chicago although seven in 10 American adults claim to be trying to eat healthier foods, almost that many – 65 percent – say calories don't always count, with about half of Americans finding nutritional value the important factor.

According to NPD Group Inc., a research firm in Port Washington, N.Y., obesity rates have held steady for four years now. Diet still will be big; as are more than 60 percent of Americans. But our interests will turn toward other health concerns. Allergies (including gluten and lactose intolerance), energy, immunity and the more general "feeling better" issues are moving up to occupy a growing portion of the megatrend.

Another study by Mintel notes about 36 million Americans claim to suffer from either a food allergy or intolerance. And a Natural Marketing Institute report finds two-thirds of baby boomers are most afraid of fatigue as they age, and nearly half are worried about diminished mental capacity.

From a food and beverage manufacturing point of view, all these wellness areas are showing some of the strongest growth potential.

Modern Biotechnology in Food: Modern biotechnology and Food Quality

In Europe, a vast diversity of high quality foods provide the carbohydrates, fats, proteins, minerals and vitamins needed in the everyday diet of consumers. At the heart of food production is biotechnology. One aspect of biotechnology which has been used for centuries is the selective breeding of crop plants and farm animals to produce improved food. Another is fermentation, in use for millennia to produce fermented foods like cheese, bread, beer, sauerkraut and sausages. The first use of gene technology two decades ago opened up the potential for many additional advances in both selective breeding and fermentation. Each specific step forward might be relatively small, but together they could add up to further improvements in the nutritional quality, appearance, flavour, convenience, cost and safety of foods.

Better Raw Materials

In improving raw food materials, many plant breeding programmes have been directed towards boosting yield or allowing more environmentally compatible agriculture by increasing the resistance of crops to viruses, pests or herbicides. Increasing yield has clear benefits in helping to feed the world's ever-increasing population and could provide cheaper food.

Plants which are resistant to attack by insect pests and diseases would need fewer pesticide applications; resistant crops such as maize, tomatoes and potatoes are already being developed. Crops have also been produced with tolerance to modern, more environmentally compatible herbicides, with the aim of achieving optimal weed control with reduced levels of herbicide. Today, there is increasing interest in improving the nutritional value, flavour and texture of raw materials. This could help encourage greater fruit and vegetable consumption in line with government guidelines on healthy nutrition.

A range of promising crop plants are being developed with:
- Improved nutritional value - Crops in development include soybeans with a higher protein content; potatoes with more nutritionally available starch and with an improved amino acid content; pulses such as beans which have been altered to produce essential amino acids; crops which produce beta-carotene, a precursor of vitamin A; and crop plants with a modified fatty acid profile. An example is a strain of oilseed rape which produces a special type of polyunsaturated fatty acid (the so-called w3-fatty acids). These have been linked to brain development and have potential in a range of speciality, clinical and infant foods.
- Better flavour - For example, types of peppers and melons with improved flavour are currently in field trials. Flavour can also be improved by enhancing the activity of plant enzymes which transform aroma precursors into flavouring compounds.
- Improved keeping properties with the aim of making transport of fresh produce easier, giving consumers access to nutritionally valuable whole foods and preventing decay, damage and loss of nutrients. Examples include the improved tomatoes now being sold in the US, and recently approved in the UK, which have been genetically altered to delay softening. Research is underway on making similar modifications to broccoli, celery, carrots, melon and raspberries. The shelf-life of some processed foods such as peanuts has also been improved by using raw materials with a modified fatty acid profile.
- Reduced levels of toxicants, allowing a wider range of plants to be used as food crops, such as the edible strain of sweet lupin which has been developed through conventional breeding techniques.

Improved Food Ingredients

Necessary changes to the key food ingredients, starches and oils, are usually made by processing. Biotechnology opens up the possibility of altering crop plants to produce exactly the type of ingredients needed:
- Starches- Plant breeders have introduced a bacterial gene into potato plants which increases the proportion of starch in the

tubers whilst reducing their water content. This means that the potatoes absorb less fat during frying, giving low-fat chips. Sweeter potatoes have also been produced which have a higher sucrose content than traditional varieties.
- Oils- Both rapeseed and sunflower are being altered to produce more stable and nutritious oils, which contain linoleic acid instead of linolenic acid and have a lower saturated fat content. Rapeseed has also been modified to produce a high-temperature frying oil low in saturated fat.

Advances in Processing and Additives

Research underway at present aims to allow the production of better food raw materials by crop plants. However, some processing steps remain essential to bridge the gap between currently-available raw materials and the desired end-product.

Traditional biotechnology has played a major role in producing fermented foods- where desirable changes are produced by the action of micro-organisms or enzymes- of which over 3,500 different types exist around the world.

In Europe and North America, bread, yoghurt and cheese are perhaps most familiar. In Africa, foods made from fermented starch crops like yams and cassava are more important, whereas in Asia, products derived from fermented soya beans or fish predominate.

Fermentation can make the food more nutritious, tastier or easier to digest, and it can enhance food safety. It also helps to preserve food and to increase its shelf-life, reducing the need for additives. Genetically improved strains of microbes can make a major contribution to these desirable properties.

For many years, a wide range of additives, processing aids and supplements have been obtained from microbial sources by fermentation. Increasingly, modern biotechnology is being used here. Products include vitamins, citric acid, natural colourings, flavourings, gums and enzymes. Gums used as low-calorie thickening agents and low-calorie sweeteners from natural ingredients are also produced using modern biotechnology. Enzymes- the naturally-occurring catalysts responsible for literally all the biochemical processes of life - are used in applications such as bakery and cheese making to improve texture, appearance and nutritional value, and to generate desirable flavours and aromas.

A second area where biotechnology has advantages is in improving the processes by which food is produced. Here, it can be used to develop mild, highly specific processes using modified micro-organisms and purer, cheaper enzyme products. These can offer better productivity, cost-effectiveness and energy efficiency than existing processes. They can produce top-quality foods with a reduced need for additives such as flavourings, and can also reduce the environmental impact of food processing.

Specific areas of food processing where advances are being seen are:

- Bread-making, for which improved strains of yeast have been developed containing genes for production of other food processing aids, such as amylases, which give improved dough. Yeast can also be used to produce a range of enzymes for use in processes such as cheese production, where introducing a copy of a calf gene has given a strain of yeast which produces the enzyme, chymosin. Previously, this enzyme could only be obtained from the stomachs of calves.
- Fruit juice production, where juice yields from apples can be improved by adding pectinase enzymes. These are produced naturally by a strain of the mould Aspergillus. The rate at which the enzymes are made can be improved by transferring the gene for pectinase from one strain of the mould into a second strain with a higher capacity for enzyme production.
- Improved quality management and food safety, through a greater understanding of micro-organisms and enzymes in food production. A range of biological tools, such as monoclonal and polyclonal antibodies, will add to this impact through their use in a range of diagnostic tests aimed at enhancing the quality and safety of products and processes. These can potentially be used to monitor the presence of additives, toxins, pesticides, micro-organisms and antibiotics, and they will give quicker, more accurate detection than traditional laboratory processes.

Health is still the biggest part of the aging trend. For every age group there's a health concern some processor is targeting. Attention is split mostly among concerns of children and seniors. By sheer numbers, the last group is headed for steady growth as a trend. Basically, like Dylan Thomas, we will "not go gentle into that good night." Besides, 60 is the new 40, mirrors be damned. If a food or beverage can boast an ingredient polyphenolics, antioxidants, omega oils – to put some bloom back on the rose, then we're going to buy it. But the aging trend involves more than ingredients and formulations. The doubling of the over-65 population by 2030 means increased need for easier-to-open containers. The logistics involved can include everything from packaging machinery redesigns to food safety concerns based on conflicting needs to creating tamper-resistant and sanitary packaging that doesn't require sophisticated kitchen tools or a magician's touch to open.

Then there's labelling. Larger print on labels for presbyopic eyes means less room for those marketing blurbs, images, slogans and serving suggestions that go so far to separate one product from its competitors.

Eat Global, Buy Local

New findings in a joint study by Mintel and the National Assn. of the Speciality Food Trade, New York, show speciality food "continues to show strong mainstream movement," and it singles out ethnic influences as part of the growth surge in the $35 billion product niche.

"American consumers are continuing to take advantage of the country's diverse cultures and offerings," said Marcia Mogelonsky, senior analyst for Mintel. "As consumer interest in new flavours and products continues to grow, so does the speciality food market. As immigrants continue to acculturate, their food traditions are becoming more mainstream. More Vietnamese, Thai and Indian flavours will continue to flourish within this category."

Globalization can be a volatile category. One example is the healthy fruit açai. Before 2000, the fruit was enjoyed primarily on its South American home turf, usually at small juice stands. In 2001, the first company to bring açai to the U.S., Sambazon, San Clemente, Calif., imported 40,000 lb. of açai pulp. In 2005, the figure was about 4 million lb., according to CEO Ryan Black, and demand is growing as fast as supply can fill it.

Globalization's real heft as a trend, though, comes through such mega issues as the drive to enter the China market. (To find out more, check out our archived web cast, "Four steps to China: One billion hungry customers await you"). And as the sleeping tiger that is India awakens, another billion potential customers wait in the wings.

Control Yourself

We're controlling portions not just for health but convenience. As a trend, convenience has been high on the list of movements to follow for years. But the two aspects merged in 2004 when Kraft Foods Inc.'s

Nabisco brand launched 100-Calorie Packs of some of its most popular cookies and crackers.

Those aren't exactly health foods. But it was a clever move, because dietitians and other nutrition experts had been clamoring about portion control for years. Although aimed largely at the restaurant industry, it was a pleasant surprise when the cry for sensible portions finally was heard by the packaged food industry.

The trend took off with such gusto Kellogg Co., Battle Creek, Mich., brought out its 90-calorie packs of Special K Snack Bites and 130-calorie Granola Munch'ems. Frito-Lay North America, Plano, Texas, joined in with 100 Calorie Mini Bites versions of its Cheetos and Doritos snacks. In the space of two short years, sales of such portioned packets passed the quarter-billion dollar mark.

Even manufacturers of healthy foods are hopping on the portion-control bandwagon. For example, Nspired Natural Foods Inc., San Leandro, Calif., just went national with its 90-calorie O'Coco's chocolate crisp packs. Vitalicious Inc., New York, offers 100-calorie, natural, vitamin-fortified brownies and muffins.

Make Room Kosher, Halal is Here

Kosher broke away from ethnic as a trend of its own with the first wave of fear over mad cow disease. Halal certification, the Muslim equivalent of kosher, is finally grabbing at the same brass ring.

The religious oversight of food encompasses food safety, health and wellness, ethnicity and spiritualism with one stamp. Or two. In the past year or so, the growth of religious adherence among boomers is seeing the combining of kosher and organic. This trendlet ("trendmerge?") is still too new to show reliable growth figures. However, companies such as Wise Kosher Natural Poultry Inc., Brooklyn, N.Y., are literally betting the farm on it.

Kosher/halal isn't just about meat, although the timing couldn't be better with meat sales jumping. The number of products with some sort of kosher certification is nearing six figures. Estimates are that three-fourths of manufactured foods and beverages have, are in the process of, or are seeking some sort of religious oversight certification.

In a 2005 Kosher Food Report by Mintel, it was reported that 21 percent of food purchasers knowingly buy some kosher products, and 28 percent of all consumers purchasing kosher products are driven by taste and quality. Consumers, it was noted, see kosher as a synonym for quality.

Consumers see such products as being "safer," "healthier" and "better for you." Vegetarians have learned to trust the pareve (neither meat nor dairy) designation on kosher products because of the strictness that permits not even a trace of dairy via ingredient, equipment or handling.

In addition to vegetarians, Hindus, Muslims, Seventh-Day Adventists and just generally watchful consumers go for kosher. Sales of these certified products top $100 billion annually and are projected to increase 14 percent in 2007.

The half-trillion-dollar food and beverage business is trend-driven, make no mistake. The value to processors is in growing product lines not only to serve but to steer these trends.

This doesn't mean every new product needs to be an organic, whole-grain, 100-calorie, ethnic-oriented and kosher snack loaded with omega 3s and anthocyanins with an easy-open top and large-print labels. But indications are such a product would sell well.

As long as we remember that at the end of the game, taste trumps all trends.

Table 1 : Top trends in food processing (and their trendlets)

Trend	Trendlets
1. Organic	Non-GMO, Fair Trade, Sustainability, Regional, Minimalism, "Natural"
2. Health and Wellness	Diabetes and Obesity, Kids' Health, Food Safety, Women's Health, Allergies and Immunity, Well-Being, Energy
3. Age Awareness	Aging, Teens, Kids
4. Portion Control	Serving Size, Convenience
5. Globalization	Ethnic Flavours, Multinational Production Regulations
6. Kosher/Halal	Food Safety, Certification and Oversight, Spiritualism

Beverages Take Two Paths

A major trendlet is bubbling up in the world of potables. Beverages have become more or less. That's not a misprint. With sales of conventional soft drinks dipping for the first time in years, the things we drink are polarizing into quaffs with "more" (concentrated meal-replacements, energy drinks and smoothies) and "less" (flavoured and enhanced waters substituting nutraceuticals and exotic fruit extracts for calories). Manufacturers with examples in the first category include such companies as Unilever/Slim-Fast, Odwalla, Naked Juice and Soyblendz. In the latter category are such no- and low-calorie refreshers as Glaceau Fruitwater, Gus Grown-up Soda and O2Go. Tea is big again, too. Thanks to green tea being recognized for its health benefits, the category is now flooding the market with tea and tea-fruit juice blends that lean toward exotic fruits to take advantage of the health and wellness trends as well as globalization's growing ethnic flavour offerings. Such combos are including tres chic fruits like açai, carambola and pomegranate.

Technological Trends and Needs

Several well-established traditional processing options are available for the preservation of food. The most widely used among them, thermal processing, provides a high degree of microbial safety. It tends, however, to degrade the quality of foods to some extent. Freezing and distribution of frozen foods is an important technology which retains very well the nutritive quality, but changes the physical state and consumes a great amount of energy. The present is a time of rapid change in the field of food technologies and the pace of change is increasing. Major motivations determining trends of development of new, emerging or future food technologies are those which signify responses of food science and industry to:

- demands of consumers due to their changing lifestyles and expectations for
- fresher, more natural foods, which
- are less severely processed (less heat- or freeze-damaged),
- contain less preservatives, or are even free from "artificial" additives;
- nutritionally more advantageous food (e.g. containing less salt, less sugar or fat);
- safer food (posing no microbiological or chemical health hazards); and at the same time are
- foods convenient to handle ([semi] prepared or ready for consumption and with a sufficient shelf-life);
- needs for less energy requirement of processing;
- the necessity for lower impact on the environment.

These - to some extent conflicting - requirements motivated the introduction of less severe or 'minimal processing' technologies such as controlled atmosphere storage of fruits and vegetables, and modified atmosphere packaging of foods, or the development of extended shelf-life refrigerated foods such as 'sous-vide' cooked products. These requirements have also resulted in growing interest in new, 'non-thermal' methods of food preservation such as

- using new 'biopreservatives' or new 'protective cultures' utilizing their antagonism against pathogenic microorganisms;
- new 'physical' technologies:
  — ionizing radiation treatment;
  — high hydrostatic pressure treatment;
  — high voltage electric field pulses.

To inactivate pathogenic and spoilage microorganisms. (Other new physical methods of antimicrobial treatment such as high intensity light pulses, "manothermosonication" (combination of pressure, heat and ultrasound), or treatment with oscillating magnetic fields (FARKAS, 1997; GOULD, 2000) are either of more limited scope, or they are not yet sufficiently ready scientifically or technically for implementation). Application of the technologies listed above offers various opportunities for mildly processed products by preserving their sensory quality, nutritional value and appearance. However, the application potential of any new technologies, which are coming from research laboratories and not 'sanctioned' by centuries of empirical use, is influenced by many factors:

— technological feasibility;
— technical possibility;
— health impact
— wholesomeness of the product;
— occupational safety;
— environmental friendliness;
— economic feasibility (including their energy demand);
— infrastructural conditions/requirements;
— investment need and availability of investment power;
— political attitude;
— social consequences;
— psychological aspects/risk-benefit perception.

Whereas the first four aspects listed can be scientifically studied, and clear-cut general answers can be given according to the status of science and technology of a given epoch, the other factors are very much interrelated and depend on local conditions.

A brief sketch of the state of the art of the new/emerging technologies can be given as follows:

"New" natural antimicrobial substances, particularly the use of bacteriocins produced by some lactic acid bacteria seem to gain a role in eliminating the risk of specific microorganisms in some foods. Some strains of lactic acid bacteria are considered as protective cultures, if they can be inoculated into certain foods such as vacuum-packaged processed meats, and assert an inhibitory effect on pathogens during storage at abuse temperature, while having negligible effect on the sensory quality of the products. Because of associated flavours that can alter the taste of food, future uses of plant-derived antimicrobials as food preservatives are not likely on their own, but as part of a preservation system.

Ionising radiation is a versatile form of processing energy used already in a wide range of non-food applications. It offers various technological benefits by reducing food losses and improving food safety. Irradiation extends to solid and semi-solid foods like meat, poultry and seafood the same benefits as thermal pasteurization provided for liquids. Radiation treatment at doses of 1.5 to 7 kGy - depending on conditions of irradiation and of the food - can effectively eliminate potentially pathogenic non-sporeforming bacteria from suspected food products without affecting nutritional and technical qualities. These bacteria include both long-time recognized hazards such as *Salmonella* spp. and *Staphylococcus aureus,* as well as emerging or "new" pathogens such as *Campylobacter* spp., *Listeria monocytogenes* or *Escherichia coli* O157:H7. In addition to control of the aforementioned bacteriological hazards, "radicidation" of perishable commodities can extend their shelf-life 2- to 3-fold, and inactivate food-borne parasites, the latter being of particular importance for developing countries. After decades of unprecedented intense and wide-ranging research efforts, food irradiation is now a well understood and controllable food processing technology supported by all relevant specialized agencies of the United Nations (the World Health Organization, the Food and Agriculture Organization and the International Atomic Energy Agency) and many national scientifically authoritative bodies in different countries. Its slow implementation can be explained by the long time, which was needed to demonstrate adequately the safety and wholesomeness of irradiated food, the lack of readily available radiation facilities and their investment-demanding character, as well as an inadequate awareness of problems, which justify the use of this technology. The safety and nutritional adequacy of irradiated food have been well established, the technology is ready to use technically, and the need to improve the microbial and parasitic safety of food became a major driving force behind the implementation of food irradiation in both developed and developing countries. When irradiation is perceived as an adjunct to, but not a replacement for, GMP, it can serve as a 'critical control point' in the Hazard Analysis Critical Control Points system of safety management, a concept becoming mandatory in more and more countries.

The advantage of *high hydrostatic pressure* (HP) treatment (up to 900 MPa [i.e. as high as 9000-times the atmospheric pressure] for several minutes) is that it treats all parts of a high moisture food equally (high isostatic pressure) and it is attractive from the point of view of food product quality. At present its application is limited to a pasteurization-like process for certain foods in which bacterial spores are not a problem (low-pH products) or which have a limited chill shelf-life. Success of its further development depends on effective control by appropriate combination treatments and additional research for careful establishment of performance criteria for reduction of the number of relevant pathogens by a required safety factor (GOULD, 2000). Due to these facts, and because of costs involved in HP pasteurization/sterilization of foodstuffs, this process will remain probably rather specialised and will only be used to preserve foods for which a premium price can be obtained.

High voltage electric field pulses (HELP) to effect non-thermal inactivation of microorganisms in foods was also explored, and these studies led to the development of prototype and industrial scale devices recently. HELP treatment is the application of pulses of very high field strength (2-5 $V.\mu m^{-1}$) for a very short time (microseconds) to foods between two electrodes. The treatment requires fairly complex electronic and fluid handling systems. Studies on combination processes have shown potentially useful synergies because "electroporated" bacterial cells become much more sensitive than untreated cells (e.g. to bacteriocins). The application of HELP will probably be limited to liquid foods or liquids containing small particulates. There are still, however, considerable knowledge gaps that will need to be addressed, and regulatory hurdles to be overcome, before commercialisation of the technology, but HELP products might be in the market place in 10 years' time.

All these technologies might be utilized even more efficiently in rational combinations with other preservative treatment(s). Among future challenges for food science in the field of more efficient improvement of microbial safety and quality maintenance of food are:
- better understanding of factors affecting the establishment, survival and growth of food-borne microorganisms by multidisciplinary research on
  — physiology of microorganisms;
  — development of predictive models for their growth, survival, inactivation or the shelf-life of foods;
  — effect of food structure on its interactions with microorganisms;
  — application of molecular biology for improved characterization, typing and detection of important microorganisms;
- better utilization of synergistic interactions of "combinations"/ "hurdle effects", but avoiding stress adaptation.

Advances in predictive microbiology will assist optimization of existing processes and formulation of alternative processes with regard to their effect on food safety. If the microbiological feasibility of new physical technologies is carefully established and, particularly, new combination approaches are developed and proven effective, the opportunities for the use of some of the new techniques are likely to expand in the future. From the food industry side, there is a need to "metabolise" the novelty of the aforementioned technologies to appreciate their potential over the hurdles that must be overcome.

Food with Improved Nutritional Value

Until recently, health aspects of food were mainly thought of as the absence of detrimental components, in particular additives and unwanted compounds. In the past years, however, it has become evident that some naturally occurring components of certain foods can help to maintain a state of well being and health through optimising our body functions or to reduce risk of chronic diseases, such as certain cancers and coronary diseases. This is particularly important in that most countries are experiencing a considerable rise in the proportion of elderly people, and proper nutrition can assist us to live not only longer, but in better health.

In addition, many technological innovations are related to the recognition of the important role of food in promoting and sustaining health, resulting in the concept of *"nutraceuticals"* and *"functional foods"*. Nutraceuticals are those food components, which play particular roles in maintaining health. They originate mainly from plants, i.e. they are *"phytochemicals"*. Functional foods are foods characterized by the presence of one or more components having beneficial physiological effect and being effective in the maintenance of good health. In one group of functional foods live bacteria (certain lactobacilli and bifidobacteria) provide the main beneficial effect. These foods are called *probiotics*.

Other functional foods contain *prebiotics,* i.e. substances that facilitate growth of the beneficial bacteria in the intestinal tract of the consumer. This interaction between food, nutrition and health is a new challenge for both food science and the food industry and it represents a positive approach to 'optimizing" nutrition. It is important, however, that claims on health-promoting effects have sufficient scientific substantiation.

It is important that health authorities should establish criteria for mandatory qualification of functional foods. Development of functional food should not be only a marketing claim, but rather a scientific challenge. The presence of a bioactive compound in a food does not necessarily ensure that it will be biologically active when it is consumed. The benefits of the "functional foods momentum" will not be realized unless scientifically sound and non-misleading messages are provided to consumers.

Food can be said to be "functional" if it is satisfactorily demonstrated to affect beneficially one or more target functions in the body beyond adequate nutritional effects, in a way which is relevant to either the state of well-being and health or the reduction of the risk of a disease. A functional food can be a natural food, a 'modified' traditional food or a 'novel food' as defined in the aforementioned EU Regulation.

The production of functional foods happens by:
- eliminating a component causing deleterious effect to the consumer (e.g. an allergenic protein);
- increasing the concentration of a natural component which induces beneficial effects;
- adding a *natural* substance which is not normally present, but for which beneficial effects have been demonstrated;
- replacing a component which causes deleterious effects by a component with beneficial effects;

- improving the bioavailability of food components with beneficial effects.

Functional foods must remain foods, they are not medicaments, but part of nutrition/diet. Because functional food development targets healthy subjects, proving their effects' statistical significance requires a special methodological approach. It will become increasingly important to:
- identify the mechanism of action of the active components in functional foods;
- clarify the impact that commercial processing and home-processing have on bioactive components within foods;
- optimize process parameters for maximal retention and increased bioavailability of beneficial compounds;
- understand the dynamic interactions that occur among the various components of not only the food but with the other constituents of the entire diet;
- investigate the interaction between diet and human intestinal microflora and its implication for health.

Transgenic Food

While consumers accept the functional food concept remarkably readily, the acceptance of novel foods as they are defined according to the EU Regulation (EC, 1997), and particularly transgenic food, is controversial.

Conventional breeding can be used to transfer genes only between sexually compatible organisms. Complementary to conventional breeding techniques, gene technology allows the transfer of genes between unrelated species. Thereby, breeding targets can also be achieved more quickly both in plant and animal breeding. This is one of the key, but hotly debated technologies of our times. Important topics that need to be addressed vary from legislation, such as labelling requirements, to safety and environmental issues. Concerns about application of this agricultural biotechnology are on the ecological impact of growing genetically modified foods, the impact of these crops on biological diversity, and on the safety of food supply, or the development of resistance by insect pests. However, the potential of the agricultural new biotechnologies is enormous also for developing countries. Therefore, questions about agricultural biotechnology must be addressed for people in both developed and developing countries, as we have to address the issue of food security for a world population of some 9000 million people in the year 2050. Furthermore, genetic engineering is not just a new technology for crop improvement, it is a powerful research tool that is helping to provide fresh and better insights of molecular mechanisms involved in biological processes.

It is forecasted that in the next decade about four dozens of agricultural crops will be genetically modified. It is estimated that already 8.3 million ha of genetically modified corn (20 % of the total cultivated area) were planted in the US in 1999, and that more than 66 million ha will be used to cultivate transgenic plants by 2005 in North America. No less remarkable are the opportunities and research results in the field of food biotechnology by improving microorganisms used in food fermentations, and on the exploitation of microorganisms for the manufacture of food ingredients.

Without venturing more deeply into this enormously complex problem-area where I have no expertise, I should like to limit myself to those aspects of transgenic food or food components which are relevant to the potential improvement of nutritional value and technological functionality of food or food components.

There are already a number of foods on the market which are produced using genetically modified organisms or containing GM ingredients e.g. chymosin used in cheese making, use of GM tomatoes in paste, and GM soybean and corn products. Thus, most of the transgenic products to date have been developed for agricultural and processing efficiency, and not yet with direct consumer benefits in mind such as improved taste and higher nutritional value. I share the views of those who consider that a "second generation" of transgenic crops should be devoted to achieve these advantageous compositional changes in order to serve the increased interest in functional foods and passing the benefit on to the consumer. Food R&D in the 21st century should also support consumer oriented product development because it has the potential to become part of the health care system. Several transgenic foods are now under development, which aim to develop properties of nutritional significance. Some examples:
- producing rape seed and corn with nutritionally more favourable oil composition (containing negligible levels of erucic acid);
- spinach and lettuce accumulating less nitrate;
- potato richer in starch, thus absorbing less fat when fried;
- certain cereals with increased lysine content;
- certain legumes with increased methionine content;

- rice not producing an allergen;
- "yellow rice" capable of synthesising beta-carotene;
- strawberry and broccoli producing higher levels of anticancer and antioxidant agents.

Safety evaluation, however, is a key issue, which must be addressed in relation to the development of novel or improved foods. Guidance on this complex topic is available from a report of a Joint FAO/WHO Consultation (1996) on "Biotechnology and Food Safety" (FAO/WHO, 1996).

Regarding safety assessment of novel food in relation to nutrition, special attention must be paid also to allergenicity of those foods which are produced using this modern biotechnology to avoid the potential appearance of a major food allergen in a product that is normally allergen free.

A summary of results of an EU project on the development of new methods for safety evaluation of transgenic food crops has been presented at the 3rd Karlsruhe Nutrition Symposium in October 1998, while the International Life Science Institute recently published "Consensus Guidelines" on the safety assessment of viable genetically modified microorganisms used in food (ILSI, 1999).

Principles for approving novel food, particularly genetically modified raw materials, were fundamentally different in North America than in the European Union, which created complications for the food industry and trade. The labelling of genetically modified organisms was not mandatory in the US, where the FDA labelling policy requires "biotech foods" to be labelled only if they are significantly altered. In response to a request from FDA for comments from the public on the Agency's current regulatory regime for food biotechnology and labelling, in formal comments submitted to FDA, the National Food Processors Association (NFPA) called, early in the year 2000, for a compulsory notification process prior to the marketing of new "biotech" food products. The producers of such foods should file with FDA summary documentation to support the determination of safety for the biotech food (NFPA, 2000). NFPA further supports the use of voluntary labelling of foods to indicate the presence or absence of bioengineered ingredients.

The European Union "Novel Food Regulation" established a system for formal, mandatory pre-market evaluation and approval for most innovative foods and food production processes, placing particular emphasis on genetically modified products. It requires additional specific labelling of "any characteristic or food property such as composition, nutritional value or nutritional effects, or intended use which renders the food no longer equivalent to its conventional counterpart." Although this EU Regulation provides broad guidance, it leaves the door wide open to interpretation and it is unlikely to stimulate European innovation and competitiveness. The recently finalised "White Paper on Food Safety" of the Commission of the European Communities states that "the Community provisions governing *novel foods* have to be tightened and streamlined", and it describes actions to this end (EC, 2000).

This is an area where the future depends on acceptance of rational, science-based weighting of risk with benefits and the provision of accurate and unbiased information to the consumer. The risk of not using technologies and of alternative methods should be also concerned. However, there is a multitude of ethical and social issues to be considered. There appears to be an even greater aversion among consumers towards the genetic engineering of animals than towards plant biotechnological programmes.

To analyse the risks of transgenic foods, the FAO/WHO Codex Alimentarius Commission established an "Ad Hoc Intergovernmental Task Force on Foods derived from Biotechnology" (FAO, 2000), which held its first meeting in Japan at the time when I was preparing this paper. This Task Force uses the concept of "substantial equivalence" established by the Organization for Economic Cooperation and Development as a central to the process of risk assessment and will review other methods for science-based risk assessment. It is expected that the above Codex Task Force will come up with a Codex' standard on transgenic foods. The "precautionary approach" adopted by the recently signed UN Protocol on Biosafety (Cartagena Protocol) can be used by governments in their risk management relevant to the above risk assessment. Identifying and labelling foods as having been derived from biotechnology can form a risk management system that will both protect the health of consumers and promote fair trade practices (FAO, 2000).

In my view, trusting science is one of the key factors for developing new technologies and evaluating new products, regardless of the technology or practices concerned. Unfortunately, however, food irradiation, biotechnology, and confidence in food safety are all media-vulnerable issues. Future developments will depend on more effective and balanced communication to the public, effectively addressing social, ethical and political issues, scientific questions and regulatory needs. It is important to understand the way that people perceive risk psychologically. Consumer acceptance of new processes is likely to be increased when a direct consumer benefit is recognized. However, huge information gaps exist among scientists and particularly between scientists and consumers. To match the promise offered by technological advances and optimize nutrition, the overcoming of barriers in psychological and - as a consequence - political feasibility is required. This needs not only research efforts but education in all stages of the food chain and in all sectors of communication, because "if we fail to train we fail to convince, if we fail to convince we fail".

Nutritional Quality

Food Composition

- *Energy* for the metabolic and physiological functions of humans is derived from the chemical energy bound in food and its macronutrient constituents, i.e. *carbohydrates, fats, proteins* and *alcohol*, which act as substrates or fuels. After food is ingested, its chemical energy is released and converted into heat, mechanical and other forms of energy
- Food is made of one or several *macronutrients*, which are carbohydrates, fats, proteins and alcohol (e.g. rice contains 0.5% fat, 7% proteins, and 80% carbohydrates). Food contains also *micronutrients*, such as minerals (e.g. iron, calcium, etc.) and vitamins (e.g. vitamin A, B1 or C).
- Other components of the food are *water*, and sometimes *antinutritional factors*.
- Values for nutrients in food can *vary widely* and depend on such things as: the growing environment, the particular variety or breed, the method of processing, the storage temperature and time etc. The values given in the food specifications section are an indication of nutritional value.

Energy Provided by Macronutrients

Energy is needed for essential body functions (such as breathing), growth (especially during childhood), and physical activities (working and playing).

Macronutrients provide different amounts of energy, expressed as kilocalories (Kcals). Fat provides approximately twice as much energy (9 kcals/g) as the same weight of protein or carbohydrate (4 kcal/g). As stated above, more carbohydrate than fat is usually eaten in developing countries and, therefore, most food energy in the diet in these countries is derived from carbohydrate sources. The relative concentration of protein and fat in the diet is important and is expressed by the percentage of energy in the diet provided by either fat or protein. For example, if a diet provides 2,000 kcal, of which 200 kcal is provided by fat, the fat-energy percentage is 10.

Energy and Protein Requirements

The total amount of energy and protein needed by different individuals varies a great deal, depending primarily on the amount of physical activity but also on age, sex, body size and, to some extent, climate. Extra energy is needed during pregnancy and lactation. An average population, made from people of all age and gender, will need 2,100 kcal per day. Energy requirement will vary with age, sex, physical activity, climate, etc.

Energy Density Concept

The quantity of energy that a child can consume each day from gruels depends on the number of meals, quantities consumed during each meal and the energy density of the gruels. However, in many societies, the mothers, involved in multiple tasks, cannot prepare gruels more than twice a day. In addition, babies cannot eat more than 30 to 40 ml of gruel per kilogram of body weight during each meal because of their reduced stomach capacity. In the case of gruels prepared from starch-based foods that have not undergone enzymatic or hydrothermal treatments (treatments involving water and temperature such as extrusion-cooking, drum-drying), the concentration of flour in the gruels is the main determinant of their energy density. The first approximation is that 4 kcal can be obtained from 1g of dry matter.

These gruels have a viscosity which increases very quickly according to their concentration in dry matter. Those who prepare these gruels are thus placed before the following dilemma: increase the proportion of flour with respect to water and obtain a gruel of very high viscosity, *difficult for children to swallow*, or prepare gruels of suitable consistency but of low energy density, and thus give more than three meals a day. There is only one way to increase the quantities of energy consumed by children, wherever gruels are not given more than three times per day, only one way: increase the energy density of the gruels. For that, the flours must undergo enzymatic and/or hydrothermal treatments which modify the physicochemical properties of the starch. These treatments break the starch macromolecules, limit their swelling during cooking, and consequently change the viscosity of the gruels. It thus becomes possible to prepare gruels of higher energy density while maintaining an appropriate consistency.

Measurement of Gruels/Porridges Consistency

When a mother prepares a gruel for her child, it is undoubtedly the desired consistency of the gruel which guides her in the choice of the flour/water proportions. However, from these proportions depends the *energy density* of the gruel and, consequently, its' *nutritional value (i.e. its nutrient density)*. Thus, the higher the relative quantity of flour, the more the gruel will be nutritious, but also the more will it be thick, until the maximum consistency threshold accepted by the mother or the child is reached. The manufacture of infant flours permitting the preparation of gruels of energy density likely to satisfy the energy needs of the child, necessitates the application of technologies that reduce viscosity: addition of amylolytic enzymes, malting of a fraction of the raw materials, *extrusion-cooking, drum-drying*, etc...

In order to study the effectiveness of these technological treatments, an evaluation of consistency is necessary. The parameter generally measured is *viscosity*. But out of the laboratory, the realisation of reliable viscosity measurements is very difficult. In addition, the absence of recommendations for the adoption of standardised conditions of measurement invokes a great disparity in the measurement methods employed. There are many viscometers of different trade marks and characteristics and from one instrument to another, the measured viscosity on the same gruel can vary up to a factor of 10.

To evaluate the consistency of gruels, it is also possible to use a *Bostwick* consistometer. This empirical measurement, relatively simple to use in an enterprise, is less well conceived than the measurement of viscosity but could be better correlated with the sensory appreciation of mothers. Research studies are currently being carried out in the Laboratory for Tropical Nutrition, IRD, in order to better understand the relationship between viscosity, *Bostwick* flow and the sensory appreciation of the consistency of gruels.

The compartment that has a guillotine-type system is filled with a given volume (100 ml) of gruel. At t=0, the gruel is released and the consistency parameter obtained corresponds to the distance covered by the gruel front after 30 seconds of flow. The only control condition of the measurement is the gruel temperature: generally, a temperature close to that at which the gruel is consumed, that is to say approximately 45 °C is adopted.

Macronutrients

Protein

Proteins are made up of 'building blocks' called *amino acids*, composed of carbon, hydrogen, oxygen and nitrogen (amino group). Proteins from different food sources contain different amounts of amino acids. Proteins from animal origin, such as meat, milk and eggs, contain all *essential amino acids* in balanced amounts. Essential amino acids are those that the body cannot make itself and must therefore be eaten. In contrast, proteins of vegetable origin (e.g., cereals and pulses) contain on their own insufficient quantities of some of the essential amino acids. By combining different foods, however (e.g., cereals with beans), adequate levels of all amino acids can be obtained without requiring protein from animal sources.

Proteins are required to build new tissue, particularly during the rapid growth period of infancy and early childhood, during pregnancy and nursing, and after infections or injuries. Excess protein is burned for energy.

The nutritional value of a protein food can be judged by its ability to provide both the quantity and number of essential amino acids needed by the body. Different food sources contain different groups of proteins, which are made up of different arrangements and amounts of amino acids. In general, proteins from animal sources are of greater nutritional value because they usually contain all the essential amino acids.

Proteins from plant sources, such as cereals and vegetables, may be deficient in one or other of the essential amino acids. For example, the proteins obtained from wheat lack adequate quantities of one essential amino acid, and those from beans are deficient in another but the combiantion of cereal and pulses will provide a balanced diet. Because the content of amino acids is different in each food, when they are eaten together they complement each other and the mixture is of higher nutritional value than the separate foods, and is as good as animal protein. It is important, that a variety of different types of protein foods are eaten.

Cooking can alter the amino-acid composition of protein and this usually results in desirable flavour and browning development. Very little nutritional value is lost.

Protein Intake

WFP recommends that 10 - 12% of the energy of the ration is derived from protein.

For instance for a daily ration of 2,100 kcal;
- 10% = 210 kcal
- 210 kcal/4 kcal/g = 52.5g of protein per day
- Maize contains 8.5g of protein/100g.
- Beans (Cowpeas) contain 23.5g of protein/100g.
- A ration made from 400g of maize (400 x 8.5% = 34g) and 100g of Cowpeas (100 x 23.5% = 23.5g) will bring 57.5g of protein.

Fat

Fat is also known as lipid and is mainly present in food in a form called 'triglycerides'. Triglycerides consist of glycerol and three ('tri') fatty acids. The fatty acids can be mainly 'saturated' (in butter) or mainly 'polyunsaturated' (in margarine). There are also monounsaturated fatty acids, which occur in quantity in the triglycerides of olive oil and peanut oil. Fat is energy dense (9 kilocalories per gram), it does not mix with water, so that the food in which it is found tends to be more energy dense because of the relative lack of water. Eating fat in reasonable quantity is good for health. It must be remembered that the quality of the fat eaten as well as its quantity is important. Fat also confers texture and flavour to food, enhancing its palatability.

Types of Fats

There are several types of fats, the main ones are called triglycerides, cholesterol esters, and phospholipids. They contain fatty acids, which may be saturated, monounsaturated or polyunsaturated. Some polyunsaturated fatty acids are essential for humans because they cannot be made in our bodies. One group of essential fatty acids comes from plant sources; these fatty acids are found in considerable quantity in polyunsaturated margarine and in vegetable oils. We need about 1 to 2 per cent of our energy to come from this type of fatty acid.

Saturated fatty acids come mainly from ruminant animals, such as sheep and cattle, and from milk and dairy products of these animals. Monounsaturated fat does not increase the blood cholesterol level and may decrease it (olive oil and peanut oil provide monounsaturated fat). Polyunsaturated fat decreases it.

Fat Intake

WFP recommends that 17% of the energy of the ration is derived from fat.

For instance for a daily ration of 2,100 kcal;
- 17% = 357 kcal
- 357 kcal/ 9 kcal/g = 39.7 g of fat per day
- Maize contains 3g of fat per 100g. Beans contains 2g of fat per 100g.
- A ration made of 400g of maize (400 x 3% = 12g), 100g of beans (100 x 2% = 2g), and 25g of oil will cover the daily needs in fat (12 + 2 + 25 = 39g fat).

Carbohydrates

Carbohydrates can be divided into three main groups:
- Sugars
- Starches
- Dietary fibers.

Sugars and starches in food are sources of energy. Cellulose and some related substances are not used by our bodies as a significant source of energy. Nevertheless, these components are very important as, together with other indigestible substances, they constitute dietary fiber.

Sugars

The main sugars in food are sucrose, glucose, fructose, maltose and lactose. Sucrose is obtained from sugar cane (or sugarbeet) and is usually called 'sugar'. Besides providing energy, sugars also produce the sensation of sweetness. Each sugar contributes the same amount of energy (kilocalories) to our diet regardless of its sweetness. Different sugars are not equally sweet and the degree of sweetness of a food is often not a good indication of the amount of sugars present. For example, maltose is only half as sweet as sucrose.

Starch

Starch is the main form of carbohydrate in our food. It is found in a variety of cereals, vegetables and fruit, but mostly in cereals, flours and legumes (beans, peas). Starchy foods are usually cooked to improve digestibility and give a more desirable texture and flavour. During the ripening of fruit, starch is changed into sugars, which give sweetness to ripe fruits. In contrast to sugars, starch is often accompanied by significant amounts of other nutrients including dietary fiber. Starch has the same energy value as sugars.

Carbohydrate Intake

There is no specific dietary requirement for carbohydrate because energy can also be derived from protein, and fat. However, a diet that does not contain carbohydrate can lead to muscle breakdown and dehydration. This can be prevented by 50 to 100 grams of carbohydrate per day, but levels above this are desirable. Sources of complex carbohydrates, such as starch, are recommended as these often also provide necessary vitamins, elements (minerals) and dietary fiber.

Dietary Fibers

Dietary fibre is a term that refers to a group of food components that pass through the stomach and small intestine undigested and reach the large intestine virtually unchanged. Most other nutrients are digested and are being used in other parts of the body by this stage. During its passage through the large intestine some components of dietary fiber are broken down to varying degrees and absorbed by the body; the remaining components are excreted in the faeces. The current attention being given to the role of dietary fiber in prevention of certain diseases is largely due to the observation that patterns of disease observed in Africa and Asia were different from those in Western countries. It was suggested that the dietary fiber content was associated with this difference. Although it is not yet proven, there is evidence to suggest that dietary fiber is beneficial for a good health.

Nutritionally Improved Food Feeds

The use of modern scientific practices such as biotechnology in agriculture has made it possible to introduce a specific characteristic in a particular grain that can improve its efficiency as a livestock feed. A wide range of options has been put forward by scientists and industry specialists as potential means of improving the nutritional composition of feed grains that would address the specific needs of different livestock industries.

In assessing research priorities in the area of feed grains quality improvement, there has been a lack of information on the economics of the various research options. In recognition of that knowledge gap, the Grains Research and Development Corporation (GRDC) funded a project, "Economic assessment of improving nutritional characteristics of feed grains (DAN331A)". The project was a collaborative one under the leadership of NSW Agriculture, involving the Australian Bureau of Agricultural and Resource Economics (ABARE) and ACE Livestock Consulting Pty Ltd. That project aimed to provide for the first time a comprehensive set of information on the value of improving different characteristics of feed grains for animal nutrition, and information on who was likely to receive the benefits of the research. The objective of the analysis undertaken in this study was to assess those potential new feeds and determine the economic merit of research to develop those feeds.

A comprehensive set of options for new feed types has been evaluated, to establish the options with the highest priorities for research. In addition, to provide a benchmark for the value of the nutritional improvements, other forms of feed grains improvement were also assessed. The options analysed are classified as follows:
- Feeds involving change in protein content
- Feeds involving change in amino acid profile
- Feeds involving improvement in feed digestibility and efficiency
- Feeds involving reduction in anti-nutritional factors
- Feeds involving increase in yield
- New crop options.

The nutritional value of each of the new options was compared to the "standard" or unimproved feed grain. In some of the options, the nutritional quality of the grain can be changed without affecting its yield, and without any change in agronomic practices or the cost of production. In others, there were associated yield changes or changes in the level of inputs that would be needed to produce the nutritionally improved feed grain. In assessing the relative benefits from alternative forms of improvement of nutrition of feed grains, the cost-reducing impacts of the different options have been analysed in a linear programming model that determines the least cost feed rations for the different livestock industries. The aggregate model considers 43 feed ingredients and estimates the least cost feed rations for the 12 livestock industries simultaneously. The cost-reduction from the new feeds was identified for each livestock industry. Economic welfare analysis was then used to estimate the size and distribution of the benefits of research from the feed grains quality-improving research between the producers (including input suppliers such as grain producers) and the consumers (including processors and final consumers) of those livestock products. The analysis also identified which of the livestock industries were likely to receive the benefits from each of the new feeds.

All of the new feeds were analysed using the aggregate feed demand model, to give a comparative analysis of all the feeds. A selected subset of the new feeds was then analysed using ABARE's regional model. That analysis allowed some of the key potential new feeds to be examined in detail, while still being comparable to the full set of options. The analysis also reveals that the aggregate national analysis provides a valuable assessment of the overall value of the new feeds.

When the feeds were analysed to assess the economic benefits, a large number of the options were found to have small or very small returns that would not justify a significant research input. The analysis reveals that there are some opportunities to improve the productivity and competitiveness of Australia's livestock industries by improving the nutritional characteristics of some feed grains. The feeds that provide the largest welfare benefits are: High oil lupins and Naked oats. The potential benefits from several other feeds are also sufficient to make them worthwhile research targets in the feed grains area, including: High oil sorghum, High protein lupins, Low arabinoxylan wheat, Hull-less barley, High oil oats, Low seed coat content barley, and High seed coat digestibility barley.

However, there are a large number of technically feasible potential new feeds that are not likely to produce sufficient benefits to make them a reasonable research target. Of the 25 feeds with improved nutritional characteristics that were analysed, 10 had total welfare benefits of less than $0.3 million per year and a further 6 less than $1.2 million per year. Given the expected research costs, probabilities of success and the time lags involved in developing these feeds by plant breeding, it is unlikely that these options could be expected to provide a satisfactory rate of return on the research funds required. Research funds used for these projects could well be applied to more productive projects. Several of those leading options for nutritional improvement had negative impacts on some livestock industries, so that none were able to provide universal benefits to all the industries included in the analysis. As a result, different livestock industries would rank the potential new feeds in different ways, often markedly differently.

An alternative would be to aim for yield improvement rather than seek to improve the nutritional quality of the feeds. That direction for research funding would provide economic benefits of similar or greater size than from nutritional improvement, and the evidence from the analysis in this study is that those benefits may well be more evenly spread across the different industries. Clearly, the selection of which new feeds to develop needs to be undertaken carefully, to ensure that scarce research and development funds are used to provide the best returns. The analysis in this study enables those feeds to be identified, so that research priorities for feed grains can be developed with improved knowledge of the economic consequences.

Nutritional Assessment Process for Biotechnology

Modern crop biotechnology uses new techniques to do what farmers and plant breeders have been doing for centuries: making small changes to the genetic make-up of plants to improve growing or eating characteristics.

The new knowledge and techniques that biotechnology has brought to plant breeding have significantly improved the precision and the speed of the breeding process, thereby significantly improving the efficiency with which scientists can develop new crop varieties for Asia. Genetically modified foods are the most studied food products ever produced, but it is sometimes difficult for consumers to understand the exhaustive checking and testing processes that are applied to biotechnology-derived foods. Improving public knowledge of the safety assessment of biotechnology- derived foods is now thought to be an even greater priority and challenge than further development of the science and technology itself.

What people in Asia think about biotechnology AFIC's own consumer research found that ordinary people feel poorly informed about this topic area. They would like to know more, but in language and terms that are understandable and acceptable to scientists and non-scientists alike. Indeed consumers have a right to know that their food is safe and nutritious. However, in order for consumer to feel confident that foods derived from modern biotechnology methods are safe to eat, they need some basic understanding of the safety tests and precautions that are applied the new varieties of crops and produce that are becoming available. New technologies, new language and knowledge As personal computers and the worldwide web emerged, a whole new family of words and phrases also emerged, which consumers have quickly adopted and become quite comfortable with. Similarly, modern biotechnology brings with it new language and new knowledge needs. One example of this is 'substantial equivalence'. This is a term unfamiliar to most non-specialists, but which is fundamental to the principle of safety assessment for genetically modified foods. The Food and Agriculture Organization (FAO) and the

World Health Organization (WHO) of the United Nations advocate the concept of 'substantial equivalence' as the most practical approach to address the safety evaluation of foods or food components derived by modern biotechnology.

In many Asian countries, the most readily understandable terms for foods derived from modern biotechnology breeding methods are food biotechnology or genetically modified foods. Abbreviations such as 'GM food' or GMOs are perceived as jargon. Such terminology may lead to confusion, miscommunication and even misinterpretation of the topic and related issues.

What is Substantial equivalence ? Substantial equivalence is based on the principle that, 'if a new food or food component is found to be substantially equivalent to an existing food or food component, it can be treated in the same manner with respect to safety'. For a foodstuff to be assessed as substantially equivalent to currently available products, the products in question is subjected to multiple tests and checks. These include molecular characterisation of the genetic modification, agronomic characterisation, nutritional assessment, toxicological assessment and safety assessment. For example, typical questions that have to be addressed are:

- Does the genetically modified food have a traditional counterpart that has a history of safe use?
- Has the concentration of any naturally occurring toxins or allergens in the food changed?
- Have the levels of key nutrients changed?
- Do new substances in the genetically modified food have a history of safe use?
- Has the food's digestibility been affected?
- Has the food been produced using accepted, established procedures?
- The overall goal of these tests is to determine whether the plant is substantially equivalent (in terms of chemical and nutritional composition and characteristics) to food derived from a conventional source that has a history of safe use.

A substantial equivalence evaluation focuses on the product rather than the process used to develop the product. If the new product is substantially equivalent to the conventional food or feed, then the product derived through biotechnology is considered to be as safe as the conventional counterpart. If the food produced using biotechnology contains a new trait, which changes the levels of nutrients or antinutrients, such as a higher level of a vitamin or a lower level of an allergen, the assessment focuses on demonstrating the safety of the new trait. Researchers must prepare comprehensive data to support the safety and holesomeness of new crop varieties developed through biotechnology. This process requires years of laboratory and field testing before a product can be brought to the market. This article is based on extracts from Food Biotechnology: a Communications Guide to Enhance Understanding. The Guide has been produced by the Asian Food Information Centre (AFIC) and the International Service for the Acquisition of Agri-biotech Applications (ISAAA). It is intended to provide anyone who needs to write or talk about biotechnology or who simply wants to understand the science and the issues related to this important topic area, with the necessary information resources.

The kit is designed to provide the most scientifically sound and up-to-date information about biotechnology products and processes, in language that both scientists and non-scientists can understand and agree upon. The guide is available in English and other languages. Countries in Asia have a long history of producing foods using biotechnology including soy sauce, tempeh and natto (fermented soybeans), belacan (fermented shrimp paste), cincaluk (fermented shrimps), budu and ngoc nam (fermented fish sauce), tapai (fermented milk), toddy (fermented young flowers of palm) and sake. Foods such as pickles, vinegar, bread, yoghurt and cheese are also the products of biotechnology.

Modernadvances in biotechnology are not a panacea for all the challenges that Asia currently faces in providing its growing population with a varied, safe and high guality food supply. Nevertheless, biotechnology could contribute to this goal, through the development of crops that can give improved yields, require less pesticides, result in less enviromental degradation, better nutritional profiles, and better keeping and eating qualities to name but name but a few of the potential benefits.

Some of the tests applied to improved crop varieties, to determine whether the plant is substantially equivalent (in terms of chemical and nutritional composition and characteristics) to food derived from a conventional source that has a history of safe use.

Safety assessment- Molecular characterisation - for new plant varieties produced through modern biotechnology, the source of the gene introduced into the plant is first identified. The transformation system used to insert the gene is defined as well as the number of copies of inserted genes and the integrity and stability of the genetic insert.

Agronomic traits - Usually the starting points for evaluating substantial equivalence. For example, in the case of potatoes, the traits commonly examined are yield, tuber size and distribution, dry matter content and disease resistance.

Nutritional assessment - Involves key nutrients including fats, proteins, carbohydrates and essential vitamins and minerals.

Toxicology assessment - Toxicants and anti-nutrients are compounds known to be naturally present in some crops that could have an impact on health if their levels increased. For example, solanine glycoalkaloids in potatoes or trypsin inhibitors in soybeans. The levels of antinutrients in crops produced through biotechnology are compared to conventionally produced varieties grown under comparable environmental and agronomic conditions.

Allergenicity - Genes from common allergenic foods are not used. Allergic responses to foods are almost always due to protein molecules in the food. Tests include examination of molecular structure, stability of protein in stomach and intestinal fluids and measurement of the amount of any new protein in the food.

The Concept of Substantial Equivalence, its Historical
Development and Current Use

For many years the practical difficulties of obtaining meaningful information on the safety of whole foods from conventional toxicology studies have been well recognized. This became particularly apparent from the vast number of animal feeding studies conducted to assess the safety of irradiated foods.

Animal studies are a major element in the safety assessment of many compounds such as pesticides, pharmaceuticals, industrial chemicals and food additives. In most cases however, the test substance is well characterised, of known purity, of no nutritional value and human exposure is generally low. It is therefore relatively straightforward to feed such compounds to animals at a range of doses, some several orders of magnitude greater than the expected human exposure levels, in order to identify any potential adverse effects of importance to humans. In this way it is possible, in most cases, to determine levels of exposure at which adverse effects are not present, and so set safe upper limits by the application of appropriate safety factors. By contrast, foods are complex mixtures of compounds characterised by wide variation in composition and nutritional value. Due to their bulk and effect on satiety they can usually only be fed to animals at low multiples of the amounts that might be present in the human diet. In addition, a key factor to consider in conducting animal studies on foods is the nutritional value and balance of the diets used, to try to avoid the induction of adverse effects which are not related directly to the material itself. Picking up any potential adverse effects and relating these conclusively to an individual characteristic of the food can therefore be extremely difficult. Another consideration in deciding the need for animal studies is whether it is appropriate to subject experimental animals to such a study if it is unlikely to give rise to meaningful information. In practice very few foods consumed today have been subject to any toxicological studies yet are generally accepted as being safe to eat. In developing a methodology for the safety assessment of new foods it was essential to establish a benchmark definition of safe food.

Development of a Safety Assessment Framework for GM Foods

Recognising that the development of GM foods was progressing rapidly the FAO and WHO convened an expert consultation in 1990 on the 'Assessment of Biotechnology in Food Production and Processing as Related to Food Safety'. The consultation recognised the limitations of traditional toxicological test methods when applied to whole foods and recommended that a more structured approach to safety assessment should be developed. The 1990 consultation identified the comparative principle whereby the food being assessed is compared with one that has an accepted level of safety, as being of considerable importance.

In 1993 the OECD published a report on the safety evaluation of foods derived by modern biotechnology. This report which was based on a number of intergovernmental consultations included a definition of safe food.

"Food is considered safe if there is reasonable certainty that no harm will result from its consumption under anticipated conditions. Historically, food prepared and used in traditional ways is considered safe on the basis of long term experience, even though it may naturally contain harmful substances. In principle, food is presumed to be safe unless a significant hazard has been identified."

This is not to say that many foods that are already widely consumed would not show adverse effects in animal studies if they could be fed at high enough doses. Equally given the many adverse effects that can be observed with existing foods it would be unreasonable to require a demonstration of absolute safety for novel foods.

The 1993 OECD report also expanded upon the comparative principle identified by the 1990 FAO/WHO consultation and formulated the concept of substantial equivalence.

The WHO and FAO refined the concept further at an expert consultation meeting held in Rome in 1996.

What is Substantial Equivalence

Substantial equivalence is not a substitute for a safety assessment, but a part of the assessment process. As such, it provides a useful framework for regulatory scientists. Underlying the concept is the requirement that any safety assessment should show that a genetically modified variety is as safe as its traditional counterparts, through a consideration of both intended and unintended effects. This involves consideration of a wide range of information, including agronomic properties, phenotypic changes and compositional data on critical nutrients and toxicants.

In the report of the 1996 expert consultation substantial equivalence was identified as being 'established by a demonstration that the characteristics assessed for the genetically modified organism, or the specific food product derived therefrom, are equivalent to the same characteristics of the conventional comparator. The levels and variation for characteristics in the genetically modified organism must be within the natural range of variation for those characteristics considered in the comparator and be based upon an appropriate analysis of data.'

Critical nutrients and toxicants are components of a particular crop known to be relevant to human health, as determined through our knowledge of the unmodified crop and its related species. Comparative assessment of these components and their potential for change as a result of genetic modification, together with a wide range of other information on agronomic, phenotypic and other properties, permits an assessment of the likelihood of unintended effects in a modified crop. For example, where the level of expression of a particular gene is altered, this is likely to be reflected in other changes in either the crop's composition or appearance.

In the application of substantial equivalence, these key components are Considered on the basis of the long history of safe use of the traditional counterpart and any differences are identified. The defined differences are then the subject of safety assessment, which can include nutritional, toxicological and immunological testing as appropriate.

The concept of substantial equivalence has been used extensively as a tool in assessing the safety of GM foods. In comparing a GM food with a conventional counterpart, consideration is given to both intentional and unintentional effects. A wide range of information is used in these comparisons ranging from agronomic data such as crop height, yield, flowering pattern, disease resistance etc. through to compositional data on key nutrients and toxicants. In this context, key nutrients are those food components which may have a major impact on the total diet and include fats, proteins and carbohydrates as well as minerals and vitamins. The comparison can result in one of three conclusions:

- The GMO or food product obtained from it is substantially equivalent to a conventional counterpart.
- The GMO or food product obtained from it is substantially equivalent to a conventional counterpart except for a few clearly defined differences.
- The GMO or food product obtained from it is not substantially equivalent to a conventional counterpart - either because the differences cannot be defined or because there is no existing counterpart to compare it with.

Where a food can be demonstrated to be substantially equivalent, it is considered to be as safe as its counterpart and no further safety assessment is required. Where there are clearly defined differences between the GM food and its conventional counterpart, the safety implications of the differences need to be fully assessed. Where a food is not substantially equivalent, it does not mean that the food is unsafe. However, there would be a need for extensive data to be provided to demonstrate its safety.

How Substantial Equivalence is Currently Used in Practice?

The concept of substantial equivalence has been integrated into safety assessment procedures used by regulatory authorities worldwide. In many countries e.g. the EU, Australia, Canada and the USA the safety assessment process has been formalised in a series of decision trees which guide regulators and potential applicants through the various stages of the comparative process.

At present each regulatory authority could, in theory, determine a unique set of components they wish to see analysed. In practice there is already broad agreement on key components, although work is progressing to develop international consensus on a core set of components on a crop by crop basis. Although the concept has been interpreted in slightly different ways by various regulatory authorities, the overall approach to the safety assessment is very similar in all countries. The area where there is perhaps the greatest scope for divergence is which of the three categories identified in paragraph 14 that a particular novel food is assigned to. This is largely down to differing interpretations as to what constitutes a difference from a conventional counterpart.

Whilst the foods derived from biotechnology which are on sale have all been assessed for safety using the substantial equivalence concept and the results endorsed by the respective Governments, there have been some published criticisms of substantial equivalence. Many of these criticisms are based on a misunderstanding of the concept. Nevertheless such criticisms provide a useful stimulus to ensure that safety assessment procedures are kept at the forefront of scientific knowledge.

The concept of substantial equivalence has been used by regulatory authorities worldwide for approximately ten years. In this time some 40 products have been assessed, the complexity of the genetic modifications has increased. For some products, particularly commodity crops such as soya, many millions of tonnes have been consumed with no evidence of any adverse health effects. The concept has stood the test of time, although it is important that it is kept under review to ensure it remains the most appropriate mechanism for assessing the nutritional and food safety implications of foods derived from biotechnology.

Chapter 4
Enzymology for the Food Science

Since the beginning of human history, man has used enzymes indirectly. The fermentation of sugar to ethanol in the preparation of beer and wine, production of vinegar by the oxidation of ethanol, curdling of milk by lactose fermentation are thousands of years old processes where catalytic activities of enzymes are responsible for chemical transformations. Probably the first application of cell free enzyme was in cheese making where rennin obtained from calf stomach was used. The protease rennin which coagulates milk protein, has been used for hundred of years in cheese preparation. The first commercial enzyme was probably reported in Germany in 1914. Use of trypsin, the protease isolated from animals, was shown to improve washing power of detergent over traditional products. Success in the formulation of improved quality of detergent triggered efforts towards the selection of proteases suitable for application in detergents.

Subsequently, a breakthrough in the commercial use of enzyme occurred with the introduction of microbial protease in washing powder at an affordable cost. The first commercial alkaline protease from *Bacillus* sp. was marketed in 1959 and production of enzyme added detergent soon became a big business within a few years. During the period when use of alkaline protease in detergent became popular, use of enzymes in food processing industries also gained momentum in parallel. Fruit clarifying enzyme, called pectinase, was used in fruit juice manufacturing units since 1930. Enzymes hydrolyzing starch into dextrin and glucose largely entered the food industry in 1960 and more or less completely replaced acid process of starch hydrolysis. Starch hydrolyzing enzymes (α-amylase and amyloglucosidase) for the production of glucose soon became the second largest used group of enzymes in industry after detergent protease.

Enzymes may be extracted from any living organism. Sources of commercial enzymes cover a wide range, from microorganisms to plants to animal sources. But for various reasons, microorganisms became the major source of enzymes. In commercial enzyme production, fungi and yeast contribute about 50%, bacteria 25%, animal 8% and plant 4% of the total. Microbes are preferred to plants and animals as they are cheap sources, their enzyme contents are predictable and growth substrates are obtained as standard raw materials. In addition, genetic engineering has opened a new era of advanced enzyme technology. Recombinant DNA technology has made it possible to obtain enzymes present in valued sources, to be synthesized in easy growing microorganisms and also to produce tailor-made enzyme proteins with desired properties as per customers' requirements. Enzymes retaining activity under extreme conditions of temperature, pH and salt concentrations, partially active in organic solvents are all becoming a reality. The prospects of the enzyme industry look very bright with increased market position for existing use, new use of known enzymes and new enzymes having novel industrial applications.

According to a recent release by Business Communication Company Inc. Study RC-147 NA, Norwalk, CT 06855, USA on "Industrial Enzyme Products, Technologies and Applications", the world wide total industrial production of enzymes was of valued at 1.5 billion US dollars in 1997 and it gained an average annual growth rate of 4.0%. Food and animal feed applications of enzymes are constantly dominating the use of industrial enzymes on a worldwide basis.

Large volumes of industrial enzymes are usually not purified and are marketed as concentrated liquid or granulated products with specified enzyme lives. Enzymes used for diagnostic or recombinant technology or in the fine chemical industries need to be highly purified products.

The industrial enzyme market is frequently segmented on the basis of applications, rather than the nature or classes of enzymes as enzymologists classify. Application sectors have been classified in different major sectors with respect to applications of enzymes.

1. Food enzymes
    a. Enzymes for starch processing (amylases for production of glucose)
    b. Sweetener production (glucose isomerase for fructose production)
    c Bakery products (xylanase, α -amylase, glucose oxidase for dough conditioning, dough quality, loaf volume)

d. Dairy product (rennin, lactase for milk coagulation and hydrolysis of lactose)
   e. Fruit juice (pectinase, cellulase, xylanase for juice clarification, juice extraction)
   f. Wine making (glucanase and papain for haze clearance)
2. Enzymes for technological applications
   a. Detergent enzymes (proteinase)
   b. Enzyme for textile (cellulase and laccase for microfibril removal and for improving brightness of cloth)
   c. Enzymes for leather processing (protease, lipase)
   d. Enzymes for paper and pulp processing (xylanase)
   e. Enzymes of analytical use such as:
      i. Uric acid by uricase
      ii. Ethyl alcohol by alcohol dehydrogenase
      iii. Ammonia by L-glutamate dehydrogenase
      iv. Cholesterol by cholesterol oxidase
      v. Glucose by glucose oxidase
      vi. Urea by urease.
3. Enzymes for animal feed
   a. Xylanase for fibre solubility
   b. Phytase for removal of phosphate
4. Enzymes for medical applications
   a. Digestive enzymes: Pancreatic enzymes, mammalian protease (pepsin) plant proteases (Bromelain, papain), fungal amylases.
   b. Enzymes with potential therapeutic applications
      i. Asparaginase and glutaminase hydrolyzing L-aspargine and L-glutamine to aspartic and glutamic acids respectively in the treatment of leukemia.
      ii. Urokinase and streptokinase (plasminogen to plasmin) for dissolving blood clot in heart attack.
      iii. Penicillinase for hydrolyzing penicillin during acute penicillin allergy.
      iv. Hyaluronidase for hydrolyzing hyaluronate in heart attack

        v. Collaginase for hydrolyzing collagen in skin cancer
        vi. Uricase for oxidizing uric acid in gout
   5. Enzymes for clinical and diagnostic applications
        a. Enzyme linked immunosorbent Assay (ELISA): Enzymes used are: peroxidase, alkaline phosphatase, S- galactosidase
        b. Enzyme multiplied immunoassay technique (EMIT): Enzymes used are lysozyme and malate dehydrogenase
        c. Enzymatic analysis of blood constituents: Glucose, uric acid, urea, cholesterol, pyruvate, lactate, triglyceride etc.

Immobilized Enzymes

Enzymes accelerate different chemical reactions with high specificity and are not permanently modified by their participation in reactions. But enzymes are costlier than chemical catalysts, in general, and cost effectiveness of enzyme-based processes could be reached by the repeated use of enzymes. But enzymes remain in solution with products and it is not possible to recover them easily from the reaction mixture.

If they are made insoluble or stationary in active forms, repeated use of an enzyme becomes possible. Immobilization is the process by which an enzyme is made insoluble or stationary with the retention of full or substantial activity. Immobilization is also localization or confinement of enzymes during a process, which permits separation of the enzyme from substrate and product for its repeated use. The use of insoluble form of an enzyme in a process offers a number of advantages such as:

   a) Repeated use of the same enzyme as far as practicable,
   b) Ability to terminate reaction at any stage by the removal of insoluble enzyme,
   c) Recovery of enzyme free product
   d) General improvement of enzyme stability.

Techniques used for the immobilization of enzyme activity may be classified as:

   a) Physical adsorption of enzyme on inert insoluble carrier
   b) Fixing of enzyme on insoluble support by covalent binding
   c) Entrapment of enzyme activity in polymerized gel
   d) Insolubilization of enzyme by cross-linking with bifunctional reagent.

However, it is understandable that some changes of physical and chemical properties of immobilized enzymes may take place because of the development of new microenvironment around enzyme by the supporting matrix.

The changes are usually expressed to various extents by the altered stability and kinetic parameters of the enzymes. Stability of the enzyme either increases or decreases on immobilization depending on the effect of the microenvironment on stability and denaturation of the enzyme.

Enzymes attached to inorganic matrices were found to be more stable than those attached to organic polymers. Specific activity of an enzyme usually decreases upon immobilization, possibly due to partial denaturation of protein depending on the process of coupling between enzyme and matrix. The presence of electric charge of the matrix affects pH optima of enzymes and the insoluble matrix limits diffusion of high molecular substrates due to steric hindrance.

In the immobilization of an enzyme, it is most important to select the method of attachment, which will not affect or interfere with the substrate-binding site of the enzyme. Considerable knowledge of the active site of the enzyme is essential and any possible interaction with binding site is avoided. Active site of the enzyme is sometimes protected during attachment and freed later.

Use of substrate or competitive inhibitor to protect the active site was found to be useful in some cases. Enzyme as biocatalyst can be immobilized using either purified/semi-purified enzyme (without undesired contaminant activities) or whole cells or sub-cellular components. Most of the enzymes used in industry are microbial extracellular enzymes, which can be isolated more easily from fermented broth as crude enzyme.

Extra cellular enzymes are generally more stable than intracellular enzymes against environmental stress. The cost of enzyme is kept low by the development of fermentation with high enzyme yielding strain. Lengthy and expensive methods of enzyme purification are avoided. Immobilization of whole cell containing the enzyme activity is highly cost effective and the process has some advantages. It preserves the natural environment of the enzyme and loss of activity of immobilized enzyme for each cycle of operation is generally lower than that of free enzyme. Whole cells also provide a number of catalysts present in the cell, if required in a process. But major disadvantages are limitations on the diffusion of substrate to and product from cell and possibility of unwanted side-reactions in presence of other enzyme in whole cell. Viable and non-viable cells are both immobilized. For single step processes, non-viable cells are used and cells are permeabilized by various physical and chemical treatments. Permeabilization causes diffusion of substrate or product through cell membrane but removes most of the small molecular weight compounds including co-factors from cells. Such non-growing cells are highly useful and economical source of intracellular enzymes for simple bioconversions requiring no regenerated co-factors for activity. A number of techniques have been developed for permeabilization of cells such as treatment of yeast or bacteria with toluene or sonication of cells etc.

There is no best-known method for the immobilization of any specific enzyme. The support, the enzyme, the substrate and technique, are all involved in the development of an effective process. The support material should be non-toxic, low cost, maximum biocatalyst loading capacity and with good flow character and operational durability.

Various Methods Used for Immobilization of Enzymes

In the process of immobilization by binding of enzyme to an inert carrier or cross-linking of enzyme to form an insoluble aggregate, the three dimensional structure of the active site, substrate and catalytic specificities of the enzyme must be kept unchanged during the process of immobilization. Chemical activities, free functional groups of the protein molecule such as α-, β-, γ-, carboxyl groups, free amino groups, thiol, hydroxyl, imidazole groups of amino acids are exploited if they are not involved in formation of catalytic site of the enzyme.

Adsorption on Solid Support

Adsorption of enzyme on solid surface (through Van der waals force or ionic interaction) has advantages because of its simplicity and non-involvement of any chemical reaction, which might affect catalytic activity of the enzyme. But this binding process is a reversible one and desorption of the enzyme becomes significant in presence of high substrate and ionic strength of the medium. Various inorganic matrixes such as alumina, activated carbon, clay, glass, controlled pore glass bead, hydroxy-apatide etc. are used in the immobilization process. Whole cells also could be immobilized using porous brick (yeast), silica (yeast), celite, diatomaceous earth Kieselguhr (mycelial cell), wood chips (yeast) etc. Hydrous metal oxides like those of titanium, zirconium, iron, tin, vanadium are capable of forming active enzyme insoluble complex.

Enzymes may be physically attached to insoluble matrix by ionic interaction. Ion exchange resins, e.g. DEAE-cellulose, DEAE-sephadex, Carboxymethylcellulose have been used as support media. However this binding is very sensitive to changes of pH and ionic strength of the medium.

Covalent Coupling of Enzyme on Activated Matrix

Covalent binding of the enzyme is more stable than adsorption attachment and enzyme is not leached out easily by changes of pH or at high salt or substrate concentrations. But the process, in general, is more drastic and immobilization is mostly associated with some loss of enzyme activity.

Among the available reactive groups of amino acids, free amino group of protein is predominantly chosen, although coupling through other functional groups is also known.

A few chemical reactions commonly used for the formation of covalent bonds between enzyme and matrices are as follows:

i) The widely used method for immobilization of enzyme on polysaccharide matrix is cynogen bromide activation of matrix. CNBr interacts with –OH groups of the polysaccharide positioned side by side, to produce immidocarbonate derivative. The activated polysaccharide interacts spontaneously with free amino group of amino acid forming covalent bond.

ii) Diazonium derivative of insoluble career (p- aminobenzyl cellulose) is used for coupling with phenolic imidazole groups of proteins.

iii) Hydrazide derivative of insoluble matrix is used for coupling

Methyl ester of carboxymethyl cellulose + Hydrazine = hydrazide derivative'! $HNO2$ '! Azide derivative + NH_2-Enzyme '! Immobilization

iv) Immobilization on synthetic matrix may be exemplified as

Polyaminopolystyrene + $CoCl_2$ '!Isocyanate derivative of polystyrene '! NH2- Enzyme (immobilization)

v) Glutaraldehyde is a useful reagent for the formation of covalent linkage between enzyme and matrix. Enzyme could be linked to aminoethyl cellulose or amino alklylated porous glass through this bifunctional reagent,

$CHO-CH_2-CH_2-CH_2-CHO$. Bis-diazobenzidine also acts as a bifunctional reagent for forming covalent linkage between enzyme and activated matrix.

Entrapment of Enzymes and Microbial Cells

Calcium Alginate

Alginic acid is a co polymer of β-mannouronic acid α-L-Glucuronic acid linked by glycosidic linkages. The polysaccharide is a constituent of marine algae. Alginate solution produces gel in presence of calcium ion. Systems reported such as: Entrapped cells of *Alcaligenes eutrophus* in calcium alginate for the production of H_2 from formate.

Production of isomaltulose from sucrose by *Erwinia rhapontici*

Photosynthesis by immobilized algae.

Carragenan

It is a polysaccharide obtained from red sea algae. It is a linear sulphated polysaccharide containing D-Galactose, 3, 6-anhydro-D-Galactose and their sulphate ester derivative. Among carrageenans, K-carrageenan is insoluble in cold water as potassium salt but is sensitive to Na^+. Some of immobilized systems are:

Production of aspartic acid by entrapped Escherichia *coli.*

Production of ethanol by entrapped Zymomonas *mobilis.*

Agarose

The polysaccharide is obtained form marine plant. It contains alternating 1, 3 linked D-Galactopyranase and 1, 4- linked 3, 6- anhydro-L- Galactopyranose (Linear polymer). Entrapment of *Escherichia coli* in agar beads for the production of hydrogen from formate.

Gelatin

Gelatin is an abundant, inexpensive and safe matrix for immobilization of microbial cells. The protein character, high hydrophilicity and good swelling properties of the matrix are highly favorable for immobilization process. After immobilization, gel is hardened by formaldehyde. A large number of enzyme activities of whole cells were immobilized (e.g. β-glucosidase, urease, invertase and acid phosphatase).

Polyacrylamide Gel

Aqueous solutions of acrylamide, N2 - N2 -ethylene bisacrylamide are polymerized in presence of suitable initiator and accelerator. Enzyme is entrapped within lattice of polyacrylamide gel. [11]β-Hydroxylation of cortexolone is done by *Curvularia lunate,* and *Aspergillus niger.*

Cross-linked Pre-polymerized Polyacrylamide

Hydrazide gel partially substituted (acyl-hydrazide) linear polyacrylamide cross linked by glyoxal.

Cephalosporin production by *Steptomyces claruligerus*.

Steroid reduction by *Mycobacterium species*.

Glucose to ethanol by *Sacchromyces cerevisiae*.

Photo Cross-linkable Resin Prepolymers and Urethane

Prepolymer methods use synthetic resin prepolymers, which are photo- cross-linkable resin, polymerize on illumination with UV light. Polyethyleneglycol dimethacrylate, polybutadiene, maleic polybutadiene and polypropylene glycol are used for the purpose. Hydrophobic photo-cross linkable prepolymer (ENT-2000, ENT-4000, ENT-6000, PEGN-1000 / 2000 / 4000) have been used for entrapping sucrase, glucose isomerase, lipase, amino acid deacylase activities.

Water miscible urethane prepolymer (PU) contains isocyanate functional groups at both termini and molecules react with each other in the presence of water. Peroxisome from methanol oxidizing yeast has been successfully immobilized by polyurethane prepolymer.

Micro Capsulation

Synthetic microcapsules are prepared by the use of hydrophilic and hydrophobic monomers. Enzyme is added with hydrophilic monomer in water and mixed with hydrophobic monomer emulsion in organic solvent. Microcapsules of 10-100 ìm diameters are formed with enzyme present in water within. Enzymes are also capsulated in liposomes. Liposomes are lipid membrane with a water droplet inside. They are usually prepared from phosphatidylcholine and cholesterol.

Cross Linking of Enzymes

Same enzyme molecules may be cross-linked by bifunctional reagents such as glutaraldehyde and bis-benzidine. Sufficient cross-linking, occurring at high enzyme concentration makes the enzyme insoluble. However the technique is not much preferred, as it requires large quantities of enzyme and insolubilization is associated with some loss of enzyme activity.

Commercial Processes in Operation

Among the large number of processes reported for the immobilization of commercial enzymes, immobilized enzymes operated successfully in large-scale commercial process are:

a) Isomerisation of glucose for fructose production.
b) Aminoacylase system for amino acid production.
c) Lactase for hydrolysis of whey lactose.

Processes for producing high fructose corn syrup using glucose isomerase use both granules or amorphous or fibrous form of immobilized enzymes with productivities ranging from 1000 - 9000 Kg of 42 % fructose syrup /kg enzyme. The different immobilization procedures used include, DEAE cellulose adsorbed enzyme (*Streptomyces* enzyme), cross-linked lysed cell (*Bacillus coagulans*), gelatin entrapped and glutaraldehyde cross-linked whole cells (*Actinoplanes*), inorganic carrier adsorbed cells (*Streptomyces spp*) etc.

Lactase hydrolyzing lactose into glucose and galactose has a number of applications, particularly in utilizing lactose present in appreciable concentration in whey. In this respect, a large number of immobilization techniques were reported from time to time. Examples include covalent attachment of lactase to glass, collagen, sepharose, entrapment of enzyme in hollow fibre, polyacrylamide gel, and cellulose acetate and in ionizing radiation induced polymers of acrylate and methyacrylate. Lactase immobilized by binding on cellulose sheets was found to be stable for several months. *Aspergillus* lactase immobilized on controlled pore glass or titanium was reported to be available at a very low cost.

Commercial process using aminoacylase system for deacylating acetyl DL– Methionine operated commercially using different immobilized systems. The enzyme attached to DEAE- sephadex by ionic interaction, enzyme covalently bound to iodoacetyl cellulose, enzymes entrapped in polyacrylamide gel were used in different commercial processes.

Production of Glucose from Starch

Starch is synthesized naturally by a variety of plants but some plants produce a high amount of starch such as corn, potato, rice, sorghum, wheat and cassava. Starch contains two types of polysaccharides both containing glucose. An unbranched single chain polymer of 500-2000 or more glucose residues, linked through α-1, 4- glucosidic linkages is called *amylose*. The other fraction called *amylopectin* is branched where branch α- 1, 4-linked glucose chains are linked through α-1, 6- glucosidic linkages to the main chain. The degree of branching in amylopectin is about one per twenty-five residues of unbranched fragment of the main chain.

Starch in general is not soluble in water. It is partially crystalline and quite compact due to intra and inter-molecular hydrogen bonding. When an aqueous suspension of starch is heated at higher temperature, hydrogen bond gradually become weaker and water molecules enter into the starch granules. This process, known as gelatinization, makes starch susceptible to the action of enzyme. Gelatinization is usually accomplished at 90-100° C in water. Addition of alkali to starch suspension and neutralization of alkali also causes gelatinization of starch at room temperature.

Corn Syrup

The knowledge that starch yields a sweet substance when heated with acid was available to chemists two centuries back. Corn emerged as the best source of starch because of its low cost, high availability and long storage life. Acid hydrolysis of cornstarch to produce 42 dextrose equivalent (DE) syrup was commercialized in early 20th century but the product could not compete with sucrose in terms of sweetening power or taste. During 1940-60, the discovery and isolation of amylases from microorganisms led to the development of a number of processes for the production of corn syrup. Various enzymatic hydrolysis protocols were introduced for the production of corn syrup in different food processing industries. Subsequently crystalline dextrose was obtained at 95% yield from cornstarch using amyloglucosidase along with bacterial and fungal α-amylases. The corn hydrolysed products known as "corn sweeteners" are sold on the basis of reducing sugar content or dextrose equivlent (DE). Higher DE value indicates higher degree of hydrolysis. The hydrolyzed product dextrose however remains contaminated with short or medium sized oligosaccharides and their presence lowers the DE of dextrose from 100 to various values. Different types of corn syrup have DE from 20 - 99.5, syrups with DE lower than 20 are usually called as malto-dextrin syrups.

Enzymatic Hydrolysis of Starch

Amylases are the enzymes, which hydrolyse starch to different extents. Enzymes hydrolyzing starch are produced by a wide variety of living beings, including humans. Human saliva and pancreatic secretions contain large amount of α- amylase for starch digestion. However major classes of α- amylases are produced by microbial fermentation. The important enzymes used in starch- saccharification process are α-amylase (EC 3.2.1.1), β- amylase (EC 3.2.1.2), amyloglucoside or glucoamyalse or γ-amylase (EC 3.2.1.3), pullulanase

(EC 3.2.1.41) and isoamylase (EC 3.2.1.68). In the hydrolysis of starch, enzymes should specifically hydrolyze both α- 1,4 and α- 1,6 linked glucose molecules in starch.

α- Amylase (α-1,4 glucan-4-glucanohydrolase)

The enzyme is an endo- glucanase, hydrolyzing α-1,4- glucosidic linkages in the interior of the starch molecule. The action of the enzyme is stopped when α- 1,6-glucosidic branch linkages are reached. The hydrolysis of amylose fraction of starch by a α- amylase results in the production of oligosaccharides or dextrins from 1,4-linked glucosidic chain in first phase. Dextrins are subsequently hydrolyzed into maltose and glucose. When amylopectin is the substrate, the hydrolysis products consist of a mixture of unbranched and branched oligosaccharides in which α-1,6-bonds are present The enzymes produced by different groups of bacteria and fungi are classified into liquefying and saccharifying amylases. Liquefying amylases quickly lower the viscosity of gelatinized starch by converting starch into dextrins but slowly hydrolyse dextrin into maltose. Saccharifying amylase, on the other hand, slowly lowers the viscosity of starch by dextrinization but quickly produces maltose and glucose from dextrins. A large number of *Bacillus* species and fungi like *Aspergillus, Penicillium, Mucor* and *Rhizopus* produce saccharifying and liquefying α-amylases. Amylases from *A.oryzae, A. niger, Bacillus amyloliquifaciens, Bacillus licheniformis* are commercially produced.

β - Amylase (α, 1,4- glucan Maltohydrolase)

This enzyme is an exo-enzyme that liberates maltose from linear non-reducing end of the polysaccharide. When β - amylase acts on amylase, it yields maltose quantitatively. But amylopectin is digested by the enzyme to yield maltose and a highly branched dextrin as the enzyme could act on α-1,4-linkages of amylopectin upto 2-3 glucose residues away from the branched 1,6-linkage. Thus, starch which contains both amylase and amylopectin, is hydrolysed by β - amylase into maltose and a highly branched core of amylopectin. Microbial producers of the enzyme are *Bacillus polymyxa, B.megaterium, Streptomyces spps, Pseudomonas spps, Rhizopus japanicus*. Bacterial enzymes are mostly active near pH 7.0 and do not require Ca^{2+} for optimal activity.

Amyloglucoisdase (α-1,4- glucan-glucohydrolases)

The enzyme amyloglucosidase or glucoamylase is an exo-enzyme that liberates α-1,4-linked glucose residues consecutively from non-reducing end of the starch molecule. The enzyme can hydrolyse terminal α-1,6-glucosidic linkages but much slower than α-1,4-linkages. Activity of the enzyme is lowest on maltose and increasingly higher with the increase of oligosaccharide sizes upto 5-6 glucose units.

Fungi are most active producers of amyloglucosidases. Enzymes are obtained from *Apergillus niger, A.oryzae, A.awamori,* and different strains of *Rhizopus.* Amyloglucosidses are frequently produced as isoenzymes by a single strain and all of them are not equally active on cornstarch.

1-6 Glucoside Splitting Enzymes

Enzymes capable of hydrolyzing α-1,6-branching of amylopectin are very important in the saccharification of starch into glucoses. Enzyme hydrolyzing amylopectin directly are pullulanase (EC 3.2.1.41) and isoamyalse (EC 3.2.1.68). Addition of pullulanases during enzymatic hydrolysis of starch by α amylase/ β amylase has a strong synergistic activity on the yield of glucose from starch. Isoamylases are produced by *Bacillus, Serratia,* and *Pseudomonas* spps. Pullulanase is capable of hydrolyzing both pectin and α -glucan microbial polysaccharides including pullulan, but isoamylase acts only on pectin. Pullulanases are produced by bacteria belonging to the genus *Aerobacter, Bacillus, Pseudomonas, Streptococcus,* etc.

The Manufacture of Dextrose from Starch

The process involves the use of thermostable amylase and amyloglucosidase (and pullulanase) for hydrolysis of starch into dextrose monomers at yields close to quantitative. Gelatinized starch is liquefied and dextrinised by thermostable α-amylase. In a typical process this step is carried out in two stages with a holding time of 5-8 minutes at a temperature of 104 °C -107 °C followed by for 90-120 minutes at 94 °C - 97 °C. The gelatinized starch slurry contains suitable does of thermostable α amylase.

The product obtained is soluble dextrin, a mixture of average DE of 10-15. To minimize colour formation due to Maillard reaction, under high temperature the protein content of the slurry is kept low and reaction is carried out at pH 6.0-6.5. The dextrin mixture is hydrolysed by amyloglucosdiase into dextrose. A number of factors determine the final yield of dextrose, such as initial DE of substrate, pH temperature, concentration of solid, amyloglucosidase dose. In a typical process 94-96 % dextrose is obtained by amyloglucosidase treatment of dextrin syrup (30-35 % dry weight) at temperature of 60 °C and pH 4-4.4 for 65-75 hrs.

Enzymatic Hydrolysis of Cellulose into Glucose

The huge plant-biomass, produced every day on earth due to photosynthesis contains cellulose as the major component. Microbial cellulose utilization, on the other hand, is responsible for largest return of the fixed carbon to the biosphere. Cellulose utilizing microorganisms present in soil and in the guts of animals play an important role in completing the carbon cycle on the earth. Cellulose degradation by microorganisms has a major role in anaerobic digestion, composting and for the supply of dietary protein by ruminants.

Cellulose is present in plant biomass in the range of approximately 35 – 50% of plant dry weight. In nature, pure cellulose is obtained from a few sources like cotton, bacterial cellulose etc. In plants, cellulose fibres are embedded in a matrix containing hemicellulose and lignin, which comprise 20 to 35% and 5 – 30% of plant dry weight. The composition of the complex termed as 'lignocellulosics' vary with plant cell type and also with the maturity of the plant. Cellulose is a linear polymer of D-glucopyranose joined together by β-1, 4-glycosidic linkages. Cellulose is synthesized as linear chains, which undergo self-assembly to produce cellulose fibres. Cellulose is initially assembled in elementary fibrils, which are further packed into larger units called micro fibrils, which are finally assembled to cellulose fibres.

Hydrogen bonds and van der waals forces act between adjacent cellulose molecules giving rise to the parallel alignment and crystalline structure of cellulose. The extensive intra- and inter-chain hydrogen bonding produces straight, stable fibres of high tensile strength.

A crystalline structure is a structure where all atoms in the molecule are present in fixed positions relative to each other. Thus crystalline cellulose prevents entry not only of enzyme but also of small molecules due to its extremely tight structure. But this crystalline structure is not uniformly distributed in fibre structure. There are many non-crystalline amorphous regions arising out of twists and other structural variations. These amorphous regions allow penetration of large molecules including cellulolytic enzymes.

Enzymatic hydrolysis of lignocellulose is more complex than that of pure cellulose. Diverse arrangement of cellulose fibres in different plant cells and presence of hemicellulose and lignin makes cellulose hydrolysis difficult.

Cellulose degrading ability is widely distributed amongst different bacterial genera and in many fungal groups. Cellulolytic bacteria include anaerobic Gram positive *Clostridium, Huminococcus,* Gram-negative *Butyrovibrio, Acetivibrio* and aerobic Gram positive *Cellulomonas, Thermobifida*.

Fungi are better known microorganisms decomposing cellulosic substrates; several species of anaerobic fungi and aerobic fungi including *Ascomycete, Basidiomycetes* and *Deuteromycetes* produce cellulolytic enzymes. Cellulolytic enzymes from the genera *Trichoderma, Aspergillus, Penicillum, Fusarium, Geotrichum etc.* have been studied in details. The hydrolysis of cellulose into its monomeric component glucose, involves the participation of a number of enzymes, collectively known as cellulolytic enzymes or cellulases as such.

During the past two decades, extensive biochemical studies have been carried out on cellulase enzymes from aerobic and anaerobic bacteria and fungi. Three major types of enzyme activities identified as components of cellulase enzyme systems are:

  i) 1,4- β-D-glucan-4-glucanohydrolase (Endoglucanase, EC 3.2.1.4)

  ii) 1,4- β -D-glucan glucanohydrolase (Exo-glucanase, EC 3.2.1.74)

  1, 4- β -D-glucan cellobiohydrolase (Exo-glucanase E.C.3.2.1.91)

  iii) β -glucosidase or β -glucoside glucohydrolase (EC 3.2.1.21).

In the hydrolytic process, the mechanism of enzyme action was suggested to be as follows.

Endoglucanases cut at internal amorphous sites of cellulose causing release of oligosaccharides of various chain lengths. Exo-glucanases act progressively on cellulose oligosaccharide chains, and liberate either glucose (glucanohydrolase) or cellobiose (cellobiohydrolase) as major products. Finally, β-glucoside hydrolyses soluble cellodextrins and cellobiose into glucose. A general property of most cellulases is their molecular structure, which consisted of a catalytic domain and a carbohydrate binding domain (CBD). CBD facilitates cellulose hydrolysis by bringing the catalytic domain of the enzyme close to the surface of insoluble cellulose substrate.

Cellulase systems are not simply mixture of these three enzymes, but an efficient combination of enzymes, which act, in a coordinated fashion to hydrolyze cellulose into glucose efficiently. Filamentous fungi and actinomycetes have the ability to penetrate cellulosic substrate by the elongation of mycelia and are capable of releasing enzyme on the substrate. These enzyme systems with or without CBM are called 'non-complexed cellulose systems'. Anaerobic bacteria, which cannot penetrate cellulosic substrate, produce 'complexed cellulase system' or cellulosome.

Non-complexed Cellulase System

The cellulase system of aerobic filamentous fungus *Trichoderma reesei* was studied most extensively, followed by that of *Fumicola insolens* and of aerobic actinomycete of genus *Cellulomonas*. *T. reesei* produces at least five endoglucanases (E.g. I-V), two exo-glucanase (CBH I and CBH II) and two β -glucosidases. Exo-glucanase II and I preferentially attack microcrystalline cellulose from reducing and non-reducing ends of the polysaccharide. They are the major components of cellulase protein of *T.reesei*. The specific role of five individual endo-glucanases is not known clearly, but synergism between *endo-* and *exo*-glucanase has been observed for each enzyme. However, cellobiose the major end product inhibits activities of both *exo-* and *endo*-glucanase. The β-glucosidase produced by *T.reesei* hydrolyses cellobiase and small oligosaccharides to glucose. Both enzymes are present mostly in the cell bound form and little in the supernatant. Cellulase system of *H. insolens* is similar to that of *T. reesei* having two cellobiohydrolases, CBH I and II and five *endo*-glucanases. Cellulase system of cellulomonas is, however, slightly different from that of *T.reesei*. It contains one exo-glucanase and six *endo*-glucanases. However, each enzyme resembles that of *T. reesei* and has CBM.

Complexed Cellulase System

Anaerobic microorganisms produce complexed cellulase system present in 'cellulosome' containing both cellulolytic and non-cellulolytic enzyme activities. Cellulosome is produced as protuberance on the cell wall of anaerobic cellulolytic bacteria like different strains of Clostridia, e.g. *C.thermocellum, C.cellulolyticum, C.cellulovorans, C.josui* etc. and rumen bacteria *Ruminococcus*.

Cellulosome contains a large non-catalytic scaffolding (CipA) protein of 197 KDa size, which is anchored to the cell wall of bacteria. About 22 catalytic proteins are also present in cellulosome. Identified activities are those of nine *endo*-glucanases, four *exo*-glucanases, five xylanases, one chitinase and one lichenase. Cellulosomes are highly glycosylated and are stable protein complexes of 2 to 16 Mda, resistant to protease action.

Acid Hydrolysis of Cellulose for Production of Glucose

The process for the production of glucose by acid hydrolysis of cellulose has many drawbacks. Since hydrolysis of cellulose is to be carried out at high temperature where damage of equipments at high acidity is quick. So cost of hydrolysis equipment is very high also the sugar is spontaneously degraded in acid environment when released.

Additional steps of acid neutralization of hydrolyzed product and purification of product from hydrolysate are also required to be included in the process. A few plants were operated in USA and Europe before World War II.

In the slow acid hydrolysis process, dilute acid was percolated down through packed bed charged with wood chips with continuous removal of sugar solution from the bottom. The reactors were heated with steam at 50 psi. The hydrolysis was continued for 2-4 hour and acid was neutralized with lime. However yield of sugar from cellulose was never higher than 40%. Since hydrolyzed liquor was a dilute sugar solution, recovery of alcohol by fermentation was also costly. Acid hydrolysis for the production of more concentrated solution was also attempted using high-pressure superheated steam. But none of the processes were found to be economic for large-scale production of directly fermentable glucose from cellulose.

Enzymatic Hydrolysis of Cellulose into Glucose

A wide variety of cellulosic substrates are easily available at a low cost, but the main problems associated with low biodegradability of lignocellulosic compounds are:

i) Cellulose is present mostly in crystalline form, which is relatively less accessible to hydrolysis.

ii) Low porosity, which does not allow the cellulolytic enzyme to gain access to the cellulose fibres.

iii) Cellulose micro fibrils are tightly surrounded by lignin, which acts as a cement between cellulose fibres. In addition, hemicellulose (xylan) and other structural polysaccharides form a physical barricade to enzyme action.

Pretreatment of Cellulosic Biomass

Milling of cellulose to 200-mesh increases enzymatic conversion of substrate to sugar by 60-70%. But milling is very energy and capital intensive. The degree of polymerization of cellulose also could be reduced by treatment with alkali, NH_3, or NO, all of which break the fibril hydrogen bond. Treatment of cellulose with ozone or cadoxen also lowers crystallinity of cellulose. High-pressure stem explosion converts the recalcitrant substrate into a hydrolysable substrate for cellulase. But all these pretreatment operations are costly and contribute largely to the cost of production of glucose from cellulosics.

In the commercial production of glucose from cellulose, cost of enzyme, however, is not favourable still today. Highest activity so far, has been reported for enzymes from *Trichoderma reesei* - Rut C-30. Cellulolytic enzymes so far obtained from other microbial sources are also not available at an economic price. Although many pilot plant studies have been carried out, enzymatic hydrolysis of cellulose has not yet been operated on a commercial scale.

Lactase in the Dairy Industry

Lactose is present in milk at about 4.7% (w/v) concentration. Unfortunately, a large fraction of world's adult population shows lactose intolerance and has difficulty in consuming milk and dairy products. Severe tissue dehydrating diarrhea and death are also reported after feeding milk to lactose intolerant children and adults suffering from protein-calorie malnutrition. The availability of low lactose-milk is very important in food-aid programme. In cheese making, the supernatant left after coagulation of milk protein, contains appreciable amount of lactose and the whey cannot be easily disposed off into the environment because of high oxygen demand. On the other hand concentration of whey to syrup causes crystallization of lactose, giving a sandy texture.

Enzyme lactase, a β-galactosidase (EC 3.2.1.23), hydrolyses lactose to glucose and galactose. The hydrolysis of lactose in the milk prevents digestive problems of lactose intolerant population. Lactose hydrolysis has more advantages, it increases sweetness of resultant milk, and it improves scoop and creaminess of ice cream, yoghurt and frozen deserts. The disposal problem of cheese-whey may be managed by the hydrolysis of lactose, when product becomes four times sweeter and whey may be concentrated to microbiologically secured syrup (70%, w/v).

Although a number of microorganisms are known to produce lactase (β-galactosidase), enzymes from *Aspergillus niger, A. oryzae, Kluveromyces fragilis and K .lactis* are supposed to be safe as they have history of safe use for human consumption.

Hydrolysis of lactose is done either by free enzymes in batch process or by immobilized enzyme or by immobilized whole cell. Several commercial immobilized systems have been developed for large-scale operation. In Italy, industrial scale milk processing technology uses fibre-entrapped yeast lactase in a batch process.

Properties of lactase (temperature and pH optima) from different microorganisms determine their applications. Fungal lactases, which work in the acidic range of pH 2.5 – 4.5, are used for acid whey hydrolysis. Yeast and bacterial lactase in neutral pH 6 – 7.5 are suitable for milk and sweet whey hydrolysis. Another important property, determining the application of the enzyme is inhibition of enzyme activity by the product galactose. Enzyme more susceptible to inhibition by galactose is only operative in a dilute solution of whey (at low lactose level) in immobilized column.

Fructose Production

Production of glucose syrup of high dextrose equivalent (DE) from corn became a very successful biotechnology by the use of amyloglucosidase in the saccharification process during 1940 – 50. But sweetness of D-glucose is less (70%) than that of sucrose on weight basis and lower solubility of glucose also posed problems. A commercial syrup containing 71% (w/v) of 97DE glucose has to be kept warm to avoid crystallization while syrup with lower concentrations of glucose is susceptible to microbial contamination. In this respect, fructose consequently received attention as being a sugar with 30% more sweetness than sucrose and two times more soluble than glucose.

Conversion of glucose into fructose requires a single isomerization step. Although chemical isomerization of glucose under alkaline conditions is possible yield is low, reaction is very slow and large numbers of undesired byproducts accumulate in the reaction mixture. Enzymatic conversion of glucose into fructose was known earlier. During glycolysis, the conversion needs participation of three enzymes: hexokinase, phosphohexoisomerase, fructose-6-phosphatase and of ATP. The whole process was commercially unviable, because of unavailability of these enzymes as commercial products and the cost of ATP. The existence of any enzyme capable for isomerizing glucose into fructose was not known before 1950. An enzyme capable of isomerizing D-xylose into D-xylulose was first reported from bacteria. The enzyme initially designated as 'xylose isomerase' was found to catalyze conversion of α-D-glucopyranose into α-D-fructofuranose in presence of cobalt ion. Subsequently, a large number of glucose isomerases were studied onwards from different microorganisms and the name 'xylose isomerase' was replaced by 'glucose isomerase' as enzyme induction did not require xylose to be present as an inducer.

Glucose isomerase (EC 5.3.1.5) catalyses the isomerisation of glucose to fructose. The reaction is reversible and a mixture of fructose and glucose is always obtained by the action of the enzyme on glucose. Most glucose isomerses used today are D-xylose isomerase (Xylose '!xylulose) which are active on D-glucose and also on D- ribose. The enzymes used in different industries usually have high temperature optima. Commercial glucose isomerase producers are *Bacillus coagulans, Streptomyces* and *Actinoplanes* species and *Flavobacterium arborescens*. Some bacteria like *Escherichia intermedia, Aerobacter aerogenes* also isomerise glucose through glucose phosphate isomerase enzyme. The commercial process for the production of glucose from fructose became possible only when enzyme was successfully immobilized. The enzymes are usually intracellular, and usually whole cells or broken cells are used as a source of the enzyme for immobilization.

In the modern process for the production of high fructose syrup, the basic process is enzymatic isomerisation of dextrose by immobilized glucose isomerase. But the process in total, includes many steps, like dextrose production (usually from corn starch), purification of dextrose syrup to make it suitable for feeding to glucose-isomerase immobilized column, isomerisation of dextrose to produce 42% fructose, refining of fructose and fractionation of the syrup for high fructose corn syrup production.

The most important factor in the use of corn dextrose syrup is the purity of the feedstock. The feedstock should have low colour, low ash content and low level of impurities such as metal ions and proteins. The refining process includes use of series of filters to remove traces of particulate matter, decolourizarion and deionization of the filtrate respectively with activated carbon and by cation-anion exchange resins. Magnesium ion is added in the feedstock, which activates glucose isomerase system. Glucose isomersase enzyme is also protected by magnesium from inhibitory action of calcium ion, if present in the feed.

Many immobilized isomerase preparations are available commercially. But cost of the immobilized enzyme is an important factor. The quality of enzyme is judged with respect to longer half – life and higher initial activity.

The conversion of glucose to fructose from 94–96% dextrose attained equilibrium at 60°C, when fructose level reached to 47-48%. But a long residence time of dextrose with enzyme is required to achieve the equilibrium. The conversion level is usually maintained at 42% of fructose with lower residence time. During the process, rate of conversion is proportional to the enzyme activity of the immobilized column. But enzyme activity decreases gradually with time during operation, which either lower fructose concentrate at fixed residence time, or increases residence time for targeted fructose concentration. In operation, a battery of columns is used in place of single one to maintain desired yield at optimum period of time. Another important feature in the process control is the operating temperature and pH. Usually process is conducted at 55°C in pH range of 7.0 – 7.5. Enzyme lives are usually between 3-4 months and columns are generally replaced twice a year.

High Fructose Corn Syrup

Commercial 42% fructose syrup produced frequently does not comply with the quality specifications demanded for the use in soft drinks. Fructose syrup containing more than 55% sucrose was essentially required for high quality soft drinks. The composition of the equilibrium mixture obtained during action of glucose isomerase, ultimately determines the percent of fructose in the reaction product.

Thermodynamics of the system favours fructose production at elevated temperature. The enzyme reactor, which could operate at 95°C, is likely to yield a reaction mixture containing 55% fructose. Water soluble organic solvents also bring similar effects on equilibrium mixture.

But none of the processes were found to be economically viable. Attempts were subsequently made to isolate fructose-rich fraction by various fractionation methods suitable for commercial exploitation.

Chromatography of fructose syrup on Zeolite or calcium salts of cation exchange resin yielded fructose-rich fraction (~90%). The fraction blended with 42% syrup yielded 55% fructose syrup suitable for use in soft drink. Other fractions eluted from ion exchange column, are rich in glucose and contain oligosaccharides. These fractions were treated with glucoamylase to obtain glucose syrup containing 20% solid.

Proteases in Food Processing

Proteases are produced commercially from plants, bacteria and fungi. Enzymes are generally grouped as alkaline, neutral and acid proteases according to their proffered pH range of activities.

Alkaline Protease

Both bacteria and fungi produce alkaline proteases. Various *Bacillus* strains (*B.amyloliquifaciens, B.subtilis, B.megaterium*), Streptomyces strains (*S.fradiae, S.griseus, S.rectus*) and fungi such as *Aspergillus niger, Aspergillus flavus* and *Aspergillus oryzae* are potential producers. Subtilisin carlsberg (*Bacillus licheniformis*) and subtilisin BPN (*B. amyloliquifaciens*) are produced in large scales. These enzymes are also called serine proteases, having serine at their active site. Proteases having higher temperature optima, good stability in presence of detergent formulations and working in alkaline range (pH 9-11) are produced commercially in large scale. Alkaline proteases of mammalian origin such as chymotrypsin and trypsin are also serine proteases, having optimum activity near pH 8.0. However these enzymes are structurally different from bacterial serine proteases.

Neutral Protease

Neutral proteases are produced by most of the bacteria and fungi, which produce alkaline proteases. Neutral proteases are relatively less rigid than alkaline proteases, being active in narrow ranges of pH and temperature. Limited uses of these enzymes in the leather and bakery industries have been reported.

Acid Proteases

We have known mammalian gastric enzyme pepsin and rennin of calf stomach for long time. Many fungi also produce acid proteases. Many enzymes are very similar to pepsin and rennin (chymosin). Enzymes produced from *Aspergillus* species, optimally acting in the pH range of 2-4 are used in many pharmaceutical preparations as digestive enzymes. Enzymes are used in industries for breaking gluten of wheat flour or hydrolyzing Soya protein.

Plant Proteases (Bromelain and Papain) in Food Processing

Plant *proteases* such as papain from papaya fruit and bromelain from pineapple have been used in food processing for centuries. These enzymes are thiol proteases having essential cysteine residue in the active site.

Papain present in leaves and unripe fruit of *Carica papaya*, was used to tenderize meats from very ancient times. Papain is commercially produced from the latex of papaya fruit. The latex is obtained by cutting the skin of unripe papaya. Fruits are normally tapped thrice at about 4 – 7 days intervals.

The flow of latex usually ceases 4 – 6 minutes after incision is made. The latex is collected and stored in polythene coated tightly filling boxes, kept under shade. During collection, precaution is taken to avoid contact of latex with heavy metals, such as iron or copper.

All containers including knives, spoons and pots used are either made of plastic or stainless steel. Papain is potentially dangerous, it damages skin on hands under prolonged contact and sometimes causes allergic reaction. The crude papain is obtained by drying the latex. The method of drying largely determines the quality of papain. Crude papain has major use in meat industry for tenderization of meat. Brewing industries in Britain also use papain for chill- proofing beer.

The latex of papaya contains many active proteinases; two major types, papain and chymopapain have been obtained in crystalline forms. Papain has a molecular weight of about 20.5 KDa and isoelectric point at pH 8.8. Papain acts over a wide pH range with optima near pH 6.0. Specificity of peptide bond hydrolysis of papain is similar to those of trypsin or chymotrypsin. It preferentially hydrolyses peptide bonds adjacent to arginine or lysine and tyrosine or phenylalanine. Papain is activated by a variety of substances like glutathione, cysteine, H_2S, HCN etc. and it binds metal ions Cu^{2+} and Hg^{2+}.

Meat of older animals can be tenderized by injecting inactive papain (oxidized disulphide form) into the jugular vein of live animal before slaughter. After slaughtering, reducing conditions developed in the meat reactivate oxidized papain, which tenderizes the meat. But later, this practice was found to be inconvenient. Action of papain in slaughtered animals could not be controlled and quality of meat was found to deteriorate with time.

Bromelain is a glycoprotein of 33 KDa molecular weight present in pineapple juice and stem tissue. It has uses in beer clarification and tenderizing meat similar to papain. However the enzyme is costlier than papain and use of bromelain is comparatively low compared to that of papain.

Protease in Cheese Making

The use of rennet in cheese making is a very old technology known to mankind for several centuries. Rennet (bovine chymosin), obtained from the fourth stomach of unweaned calves has been used for the production of cheese as a milk-clotting agent. Rennet hydrolyses specific peptide linkage between phenylalanine and methionine residues of ɣ -casein protein present in milk. The κ-casein maintains the colloidal character of milk. In the protein molecule, N-terminal hydrophobic domain of ɣ-casein remains associated with insoluble α- and β-casein while negatively charged C-terminal interacts with water and all these interactions prevent casein micelles from growing large.

Action of rennet hydrolyzing phenylalanine-methionine bond, separates the two domains with the release of hydrophobic para-κ-casein. Termination of the protective action of κ -casein causes coagulation milk to form curd. The coagulated protein is compressed and recovered as cheese. The enzymatic process is temperature dependent and requires presence of calcium ions.

An efficient transformation of milk into gel depends on both the factors. Calf rennet, which contains chymosin and a small amount of pepsin, is a relatively expensive enzyme and has limited supply. Extensive research has been carried out to obtain rennet-like proteases from microorganisms. The major utility of rennet in cheese manufacturing is its proteolytic mode of action and its easy thermal inactivation during ripening of cheese.

Presence of any residual protease was found to cause development of unwanted bitter off-flavours in cheese after ripening. Chymosin is a relatively unstable enzyme. Little activity of the enzyme remains during ripening of cheese. Many microorganisms are known to produce rennet-like proteases. *Rhizomucor pusillus, R. miehei, Aspergillus oryzae, Irpex latis* enzymes are used in cheese manufacture. Milk clotting enzyme from *R. pusillus* is obtained by the solid fermentation of the fungus on wheat bran. Enzymes from *R. michei* and *Endothis parasitica* are obtained from submerged fermentation.

Microbial enzymes with lower thermostability are more suitable for cheese making. Loss of protease activity after milk clotting similar to chymosin, is appreciated. This protease is not likely to develop off-smell of cheese during ripening. An interesting example of chemical modification of the protease, successfully lowering its thermo stability was reported.

Enzyme from *M. miehei* was treated with oxidizing agents like par acids or hydrogen peroxide to lower heat stability. The oxidation converts methionine residue of *M. miehei* enzyme to methionine sulphoxide and thermo stability of the enzyme was lowered by 10°C.

The treated enzyme was found to be very similar to chymosin, producing no off-flavour during ripening of cheese. Attempts were also made to clone chymosin into *Escherichia coli* or *Saccharomyces cerevisiae*. The calf pro-chymosin gene was expressed in *E. coli* as insoluble inclusion bodies from which prochymosin was recovered after denaturation. Enzyme produced by *Saccharomyces cerevisiae* containing cloned prochymosin gene mostly remained bound with cellular debris with only 20% in the soluble supernatant. Later chymosin was successfully cloned and expressed in *Kluveromyces lactis* and large-scale production of the enzyme was developed.

Microbial rennets from various microorganisms are marketed under different trade names. Although proteolytic activities of different fungal rennets are considerably different from that of chymosin, but acceptable cheeses are produced by the use of these enzymes.

Other Minor Uses

Heat labile fungal proteases are used in baking processes to hydrolyze gluten (protein fraction) present in flour. Weak-gluten flour is favored for biscuit manufacturing and unavailability of flour of that quality is managed by the addition of protease in dough prepared from high-gluten flour. Proteases are also used for the recovery of protein from parts (bone) of animals and fish. Controlled hydrolysis of Soya protein by proteases is carried out to produce verities of hydrolysates with different new properties.

Enzymes in Leather Industry

One of the oldest industrial uses of enzyme activity is in leather processing. Before raw hide is transformed into leather, it undergoes a series of operations whereby leather making protein collagen present in hides and skins is freed or partially freed from non-collagenous constituents. Hides and skins contain fat as well as globular proteins, *viz.* albumin, globulin, mucoids and fibrous protein such as elastin, keratin and reticulin between collagen fibres. In industry, raw material is processed through a series of operations including soaking, liming, dehairing, deliming, bating, degreasing and pickling. In pre-tanning operations, skins and hides are subjected to a water soak, which cleans the raw material and loosens the hair. The conventional and most widely used method for dehairing is the treatment of soaked raw material with lime and sodium sulphide. Subsequent deliming is also done to remove adsorbed lime from the hide. Fat present in skins is usually removed with degreasing agents such as soluble lime soap, kerosene, chlorinated hydrocarbons or spirit. In the traditional processing, a large numbers of pollution causing chemicals, lime, sodium sulphide, and solvents are released in effluents, which are toxic and cause environmental pollution.

In addition of chemical treatment, an enzymatic treatment, known, as 'bating' is an essential step to obtain optimum results. During bating, the scud is loosened and many unwanted proteins are removed. Bating makes the grain surface of the finished leather clean, fine and glossy. The traditional practice of bating is an unhygienic process where uncontrolled fermentation of leather is conducted with manures of dog, pigeon or hen as sources of microorganisms. This process gives a desired character to the finished leather. No chemical process has been developed which can substitute the fermentation.

The leather industry has a major problem regarding industrial pollution due to the use of huge amounts of toxic chemicals and biological pollution by addition of unknown microbial load to the environment. Use of enzyme in pre-tanning processes appeared to be a viable alternative technology where pollution problems resulting from tannery effluent could be significantly regulated or restricted.

Enzymes in Pre-tanning

Proteolytic enzyme is the most important enzyme used in pre-tanning process. Enzymes from plant, animals and microbial sources were known to leather industry for a long time. But development of enzyme-based process became bright with the success of production of enzymes at commercial level and availability of enzymatic formulations at cheap rate.

Animal proteases and microbial proteases from bacteria and fungi are used in leather industry. The physicochemical properties of the enzyme such as substrate specificity, temperature and pH stability and pH activity range are very important factors in the application of enzymes in different steps of pre-tanning operations. Microbial enzymes appeared to be ideal source of the proteases. Enzymes with wide pH range of activities could be produced economically from different microbes. Now various commercial enzyme preparations are available for use in leather industry.

Enzymes in Soaking

Proteolytic enzyme combinations (*Aspergillus parasiticus, Aspergillus flavus, Bacillus subtilis, and Aspergillus awamori*) active in natural or alkaline pH ranges are usually used.

Enzymes in Dehairing

Use of enzyme in dehairing is highly desired in leather procesing. It would eliminate the use of sodium sulphide, one of the most toxic chemicals with obnoxious odour. A large number of proteases from *A. flavus, A fumigatus, A. chraceus, A. effuses, Bacillus sp., Streptomyces sp.* have dehairing or hair loosening effects. Potential use of specific protease keratinase from *Streptomyces fradiae*, for dehairing was also indicated. Enzymes from fungal or bacterial sources are allowed to act at pH 10.0 for about 12 – 16 hours, and hair is removed by mechanical means.

Enzymes in Bating

The process of bating is a method for softening hides by treating them in a warm infusion of animal dung. Deliming and proteolytic actions take place simultaneously in bating. A 'bate' usually contains a proteolytic enzyme, a carrier like wood flour and deliming agents like NH_4Cl or $(NH_4)_2SO_4$. Pancreatic enzymes, bacterial and fungal proteases of neutral and alkaline types are used in bating.

Proteases in Detergent Formulation

Protease used in detergent has the largest single market accounting for 25-30% of total enzyme sales. Proteases added to detergents remove protein from clothes soiled with blood, milk and other proteinaceous materials and thus work more efficiently than non- enzyme detergents. The idea of using protease in detergent goes back to the use of pancreatic extract by Roehm in 1913. But economic and technical successes were only achieved in 1960 with the availability of enzymes from bacteria. The basic requirements for detergent proteases are:

i) Availability at low cost from safe microbial source

ii) Capable of working in high alkaline pH of common detergent solution

iii) Sufficiently thermostable with higher temperature optimum

iv) Low or no allergic response for topical use.

Interestingly the enzymes were produced from *Bacillus spps* from the beginning and alkaline proteins from these species represent the lead molecule for the protease group "Subtilisins'. Although subtilisins are classifed as serine proteases by their catalytic mechanism, but they clearly constitute a separate group in terms of their amino acid sequence and three-dimensional structure. Mainly two species, namely *Bacillus licheniformis* and *Bacillus amylolquifaciens* are large scale commercial detergent protease producers in the global market. In addition enzymes from other strains such as *B. clausit, B.lentus, B. alkalophilus, B. halodurans* etc. are used in detergents worldwide.

All these enzymes are fairly non-specific endo proteases, preferentially hydrolyzing the carboxylic side of hydrophobic amino acid residue, but are also capable of hydrolyzing other peptide bonds. They quickly convert protein substrate into small readily soluble peptides. Detergent proteases under different trade names (Alcalase, Savinase, Esperase, Maxatase etc.) are produced commercially. The enzymes are active over a wide pH range up to pH 12 and temperature up to 60°C.

In the fermentation of subtilisins, a large number of variables are manipulated to obtain high yield at low cost. Industrial production processes normally run as large scale using batch fermentations at a high cell density. The composition of fermentation media and downstream processing of the enzyme are not available in details because of company secrecy. Recent reports available for the production of bacillus subtilisins, indicated yields of enzyme in the range 20-28 gm/L of fermentation broth.

Protein engineering with subtilisin has been tried over a long period to obtain new desired quality of subtilisin. Site directed mutagenesis with the replacement of a single amino acid residue adjacent to active serine residue resulted in the production of enzyme stable in presence of hydrogen peroxide.

Hydrogen peroxide and peroxy acids are commonly used in detergents as bleaching agents. Extensive replacement of amino acid of subtilisin BPN (*B. licheniformis*) has been patented to obtain improved properties of the protease.

Earlier attempt at the production of detergent protease at commercial scale had problems due to the cases of development of allergic reaction of the workers by the enzyme dust. This was solved by the development of enzyme granules of average 0.5mm diameter.

In this process enzyme in a mixture of inorganic salt and sugar as preservative, is present in an inner core with a protective colloid. This core is coated with polymers made from polyethylene glycol along with a hydrophilic binder. The granulated enzyme was found to be more resistant toward various additives like peroxide, optical brightener, and detergent present in the formulation.

Medical Applications of Enzymes

Therapeutically used enzymes are required in relatively small amounts, but at a very high degree of purity. Many of the enzymes are used topically and orally. A few are injected into blood circulation particularly in the treatment of life threatening disorders like cancer or heart attack.

In general enzymes are foreign proteins, antigenic in nature and elicit an immune response causing severe allergic reaction, particularly when used for longer periods. Antigenic property is modified in several ways, either by covalent modification or entrapment of enzyme within artificial liposomes, synthetic micro spheres and erythrocyte ghosts. The cost of the enzyme is sometimes found to be very high. Urokinase used for dissolving internal blood clots is prepared from human urine and is very costly compared to microbial enzymes. However genetic engineering is an useful technology, which is exploited for the production of the enzymes by microorganisms.

Digestive Enzymes

Aging causes many individuals to suffer from the problem of indigestion. It has been established that after 40 years of age there is a gradual decrease in the production of digestive enzymes. In addition, poor eating habits like less chewing, late eating in the day and habits like excessive consumption of alcohol, refined carbohydrate and fat also cause inadequate production of digestive enzymes. All these problems could be minimized by the intake of digestive enzymes to improve the efficiency of digestion.

Pancreatic enzyme supplement derived from pork pancreas, is an excellent source of digestive enzymes, provides proteases, amylases and lipase. Although the enzyme supplement has a good record of success, its use has some limitations. Pancreatic enzymes are sensitive to low pH values and are destroyed by pepsin secreted from stomach.

A pH sensitive coating has been applied on several pancreatic enzyme preparations, which dissolve only above pH 5.5-6.0. Thus enzyme activities are protected in low pH condition of the stomach. Supplementation of microbial enzymes particularly those produced by different species of *Aspergillus* has many advantages. They provide a large number of digestive enzymes such as proteases, amylase, lipase, lactases, maltase and invertase. They are normally stable over a wide range of pH compared to animal enzymes. They are active and functional for longer distances in the digestive tract. However microbial enzyme should be free from microbial residue and should be a clean and pure product.

Asparaginases

Asparaginase is an important chemotherapeutic agent used in the treatment of acute lymphoblastic leukemia and other lymphoid malignancies. The treatment is based on the fact that tumor cells are deficient in aspartate-ammonia ligase activity, which limits their ability to synthesize the normally non-essential amino acid L-Asparagine.

During growth of tumor cells, L-Asparagine is drawn from body fluids. The enzyme is administered intravenously and it lowers aspargine levels in the blood stream. Asparginase does not affect normal cells as they produce sufficient L-Asparagine for their function. But concentration of extra cellular L-Asparagine is sufficiently lowered by Asparaginase, which restricts growth of tumor cells due to asparagine starvation. The major limitation in the asparginase therapy is the development of clinical hypersensitivity. Chemically modified

Asparginase covalently bonded to monomethoxy polyethylene glycol has been found to be useful for the treatment of patients who are sensitive to the native enzyme.

Native Asparaginases, obtained from *Escherichia coli* or *Erwinia chrysanthemi,* are used either alone or in combination therapy for treatment.

Sports Medicine

In sports medicine, therapeutic effects of proteases have been observed. Proteases support, enhance and regulate circulation and the immune system of the individuals, which helps to optimize workouts and muscle maintenance. Treatment of sports injuries with protease has shown good response in the recovery from inflammation and in speedy healing of bruises and swelling. It was reported that proteolytic enzymes obtained from plant sources are more effective than animal enzymes in wound healing of plastic surgery patients.

Streptokinase and Urokinase

In blood plasma, plasmin is an important serine protease, involved in dissolving fibrin clots. It is present in circulation as plasminogen which is activated to plasmin by a number of factors in tissue including tissue plasminogen activator, urokinase plasminogen activator etc.

Streptokinase is an extracelullar enzyme produced by β-hemolytic *Streptococcus*. It can exert fibrinolytic effects through activation of plasminogen. Plasmin in turn attacks fibrin to degrade it into soluble products. Streptokinase is used as an effective and cheap clot dissolving enzyme drug during myocardial infarction and pulmonary embolism. It is given intravenously after the onset of a heart attack to dissolve blood clots in the arteries of the heart wall. Streptokinase is a single chain 47kDa protein secreted by various strains of *Streptococci*.

However, the enzyme obtained from *Streptococci* remains contaminated with a number of unwanted enzymes as present in the extra cellular medium with streptokinase. Streptokinase being a bacterial protein creates immunological problems (allergic reaction) when present in the circulation. Genetically engineered *E.coli* producing a large amount of intracellular streptokinase has been obtained. Attempt has also been made to obtain streptokinase from safer strains of *Bacillus subtilis* by genetic engineering.

Urokinase is also a medically important enzyme used as an anti blood-clotting drug for treatment of heart patients. The enzyme of mammalian origin is isolated from urine. The cost of purification of enzyme is subsequently much higher compared to that of streptokinase. The possibility for the commercial production of the enzyme from human kidney cell line in hollow fibre reactor is under trial.

Glucose Oxidase Electrode

Glucose oxidase (EC 1.1.34) catalyses the oxidation of β-D glucose to D glucono- 1-5 lactone and hydrogen peroxide when molecular oxygen acts as electron acceptor. D-glucono- 1, 5 lactone, however, hydrolyses spontaneously to gluconic acid. Although the enzyme is present in many fungi but that obtained from *Aspergillus niger* has been studied extensively. The enzyme is a dimeric 160KDa protein with one molecule of flavin adenine dinucleotide (FAD) present as cofactor per monomer. FAD is not bound covalently with the apo-enzyme; it can be extracted from enzyme under mild conditions and can be added to apo-enzyme to restore catalytic activity.

The enzyme exhibits a high degree of specificity for β –D-glucose. Michaelis constant of glucose oxidase with glucose is 20mM and with dioxygen is 1.25 mM. 2-deoxy-D-glucose, D-mannose and D-fructose are also oxidized, but at a much slower rates. Glucose oxidase can transfer electron to many artificial electron acceptors other than molecular oxygen. This activity has been used widely for the quantification of serum glucose by colorometric method. In the presence of peroxidase and an electron acceptor dye, oxidation of glucose is estimated by the development colour of reduced dye.

Estimation of glucose is very important in a number of diverse situations. It is important in food industry, in fermentation processes for on line monitoring and most importantly as a clinical indicator of diabetics. Development of a portable easy analytical technique which can give immediate and reliable on-site results, is highly desired for the estimation of glucose.

Biosensor has been developed as the more convenient portable analytical tool for the purposes stated above. A biosensor, in general, is a self contained integrated device capable of providing specific quantitative and semi-quantative analytical informations using a biological recognition signal.. A catalytic biosensor (e.g. Glucose oxidase) is a kinetic device that measures steady state concentration of a transducer-detectable species formed or lost due to a biocatalytic reaction. The development of glucose biosensor became possible, as enthalpy change associated with the oxidation of glucose-by-glucose oxidase is large enough to be detected by a thermistor. It has been shown by the use of radiolabelled dioxygen that oxygen in hydrogen peroxide is derived from dissolved oxygen but not from water molecule. In principle, the of estimation of glucose may be done on with respect to:

i) pH change due to acid production
ii) Oxygen consumption
iii) Hydrogen peroxide production.

In the amperometric estimation based on oxygen consumption, the biosensor is consisted of a platinum electrode where oxygen is reduced with reference to Ag/AgCl reference electrode.

When a potential (-0.6Volt) relative to Ag/AgCl is applied to platinum electrode a current proportional to the concentration of dissolved oxygen is produced. The rate of electrochemical reduction is dependent of oxygen concentration. Depletion of oxygen concentration by the action of glucose oxidase, is detected by the reduction of current between the electrodes. The biosensor is consisted of an electrode compartment containing a platinum cathode and an annular silver electrode, connected through a saturated solution of KCl. The electrode compartment is kept separated from enzyme compartment by a semipermeable thin plastic membrane permeable to oxygen only. This compartment contained immobilized glucose oxidase separated from test solution by a membrane permeable to sustrate and products only. Biosensors of diameters from 1cm to 0.25 mm are commercially available.

The major problems of these biosensors are their dependence on dissolved oxygen concentration. Some mediator (Ferrocinium ion) which transfer electron directly to electrode is also used.

The biosensor used for the determination of unknown glucose concentration is made with glucose oxidase immobilized on the surface of polarographic oxygen electrode. In a typical process glucose oxidase is immobilized on a membrane by the use of bifunctional crosslinking reagent, glutaraldehyde. Biosensor based on glucose coating with glucose oxidase – immobilized gelatin was also developed. Enzyme has been immobilized onto gelatin by cross-linking with Chromium III acetate. The electrode has a response time of 60 seconds with detection limit of 0.25mM glucose. The accuracy was maintained till two months.

Chapter 5

Microbiology used in Wine Production

Wine is Fermented Grape Juice

Wine can be made from grapes, fruits, berries etc. Most wine, though, is made from grapes. And no matter what the wine is made from, there must be fermentation, that is, that sugar be transformed into alcohol. If the amount of alcohol is relatively low, the result is wine. If it is high, the result is a "distilled liquor," something like gin or vodka.

There are red wines, pink wines (also known as "rose" or sometimes "blush") and white wines. Red wine result when the crushed grape skin pulp and seeds of purple or red varieties are allowed to remain with juice during fermentation periods.

Pink/rose wine can be produced by removing the non-juice pumace from the must during fermentation.

White wines can be made from pigmented grapes by removal of skins, pulp and seeds before juice fermentation.

Wines might be "fortified," "sparkling," or "table."

In fortified wines, brandy is added to make the alcohol content higher (around 14 to 30 percent). These are less perishable and may be stable without pasteurization. Wines are termed still or sparkling depending upon the amount of CO_2 they contain. The carbon dioxide may be formed naturally during fermentation or may be added artificially.

Both table and sparkling wines tend to have alcohol contents between 7 and 14 percent.

Sparkling wines are the ones with bubbles (greater CO_2), like Champagne.

Table wine (which can also be called "still") are the most "natural". The alcohol concentration itself is not sufficient to preserve natural wines, they are pasteurized. The term light wine is also used to describe wine having alcohol content from 5- 10 %.

How is Wine Made?

Growing Grapes

Grapes grow on vines. There are many different types of grapes, but the best wine grape is the European Vitis vinifera. It is considered optimal because it has the right balance of sugar and acid to create a good fermented wine without the addition of sugar or water.

Harvest

Weather is a major factor is determining whether a year is going to be a "good vintage" (or "year"). For example, was there enough heat during the growing season to lead to enough sugar? At harvest time, the short-term effects of weather are quite important. To produce great wine, the fruit should have a high (but not overly high) sugar content ("brix"). Think of raisins.

As the fruit dries, the water evaporates. What is left is the sugary fruit. If it rains just at the point the wine grapes are ready, and before the grapes can be harvested, the additional water will cause the water level to increase, and the brix will go down. Not good. (You might ask, why not just add some sugar in the wine making process? Some do. Also considered "not good.")

Every year the wine grape grower plays a game of chance and must decide when to harvest. Simplistically, if you knew it wasn't going to rain, you would just test the brix until it was just right, then harvest. If you harvest too soon, you will probably end up getting a wine too low in alcohol content (there won't have been enough sugar to convert to alcohol). These wines will be "thin." If you delay harvest, there may be too much sugar, which leads to too low acid content. This also affects the taste (and the aging possibilities) of the wine.

Initial Processing of the Grape Juice

Grapes can (and might still) be crushed by stomping on them with your feet in a big vat. But a more practical way is to use a machine which does the job (and at the same time, removes the stems).

What you get may or may not get immediately separated. Skin and seeds might immediately be removed from the juice. Separation may not immediately occur (especially for red wines), since skins and stems are an important source of "tannins" which affect wine's taste and maturity through aging. The skins also determine the colour of

Microbiology used in Wine Production

the wine. Maceration (the time spent while skins and seeds are left with the juice) will go on for a few hours or a few weeks. Pressing will then occur. One way to press the grapes is to use a "bladder press," a large cylindrical container that contains bags that are inflated and deflated several times, each time gently squeezing the grapes until all the juice has run free, leaving behind the rest of the grapes. You can also separate solids from juice through the use of a centrifuge.

Crushing and Primary Fermentation

Crushing is the process of gently squeezing the berries and breaking the skins to start to liberate the contents of the berries. Destemming is the process of removing the grapes from the rachis (the stem which holds the grapes). In traditional and smaller-scale wine making, the harvested grapes are sometimes crushed by trampling them barefoot or by the use of inexpensive small scale crushers. These can also destem at the same time. However, in larger wineries, a mechanical crusher/destemmer is used. The decision about desteming is different for red and white wine making. Generally when making white wine the fruit is only crushed, the stems are then placed in the press with the berries. The presence of stems in the mix facilitates pressing by allowing juice to flow past flattened skins. These accumulate at the edge of the press. For red winemaking, stems of the grapes are usually removed before fermentation since the stems have a relatively high tannin content; in addition to tannin they can also give the wine a vegetal aroma (due to extraction of 2-methoxy-3-isopropylpyrazine which has an aroma reminiscent of green bell peppers.) On occasion, the winemaker may decide to leave them in if the grapes themselves contain less tannin than desired. This is more acceptable if the stems have 'ripened' and started to turn brown.

If increased skin extraction is desired, a winemaker might choose to crush the grapes after destemming. Removal of stems first means no stem tannin can be extracted. In these cases the grapes pass between two rollers which squeeze the grapes enough to separate the skin and pulp, but not so much as to cause excessive shearing or tearing of the skin tissues. In some cases, notably with "delicate" red varietals such as Pinot noir or Syrah, all or part of the grapes might be left uncrushed (called "whole berry") to encourage the retention of fruity aromas through partial carbonic maceration.

Most red wines derive their colour from grape skins (the exception being varieties or hybrids of non-vinifera vines which contain juice pigmented with the dark Malvidin 3,5-diglucoside anthocyanin) and therefore contact between the juice and skins is essential for colour extraction. Red wines are produced by destemming and crushing the grapes into a tank and leaving the skins in contact with the juice throughout the fermentation (maceration). It is possible to produce white (colourless) wines from red grapes by the fastidious pressing of uncrushed fruit. This minimizes contact between grape juice and skins (as in the making of *Blanc de noirs* sparkling wine, which is derived from Pinot noir, a red vinifera grape.)

Most white wines are processed without destemming or crushing and are transferred from picking bins directly to the press. This is to avoid any extraction of tannin from either the skins or grapeseeds, as well as maintaining proper juice flow through a matrix of grape clusters rather than loose berries. In some circumstances winemakers choose to crush white grapes for a short period of skin contact, usually for three to 24 hours. This serves to extract flavour and tannin from the skins (the tannin being extracted to encourage protein precipitation without excessive Bentonite addition) as well as Potassium ions, which participate in bitartrate precipitation (cream of tartar). It also results in an increase in the pH of the juice which may be desirable for overly acidic grapes. This was a practice more common in the 1970s than today, though still practiced by some Sauvignon blanc and Chardonnay producers in California.

In the case of rose wines, the fruit is crushed and the dark skins are left in contact with the juice just long enough to extract the colour that the winemaker desires. The must is then pressed, and fermentation continues as if the wine maker was making a white wine.

Yeast is normally already present on the grapes, often visible as a powdery appearance of the grapes. The fermentation can be done with this natural yeast, but since this can give unpredictable results depending on the exact types of yeast that are present, cultured yeast is often added to the must. One of the main problems with the use of wild ferments is the failure for the fermentation to go to completion, that is some sugar remains unfermented. This can make the wine sweet when a dry wine is desired. Frequently wild ferments lead to the production of unpleasant acetic acid (vinegar) production as a by product.

During the primary fermentation, the yeast cells feed on the sugars in the must and multiply, producing carbon dioxide gas and alcohol. The temperature during the fermentation affects both the taste of the end product, as well as the speed of the fermentation. For red wines, the temperature is typically 22 to 25 °C, and for white wines 15 to 18 °C. For every gram of sugar that is converted, about half a gram of alcohol is produced, so to achieve a 12% alcohol concentration, the must should contain about 24% sugars. The sugar percentage of the must is calculated from the measured density, the must weight, with the help of a specialized type of hydrometer called a saccharometer. If the sugar content of the grapes is too low to obtain the desired alcohol percentage, sugar can be added (chaptalization). In commercial winemaking, chaptalization is subject to local regulations.

During or after the alcoholic fermentation, malolactic fermentation can also take place, during which specific strains of bacteria convert malic acid into the milder lactic acid. This fermentation is often initiated by inoculation with desired bacteria.

Pressing

Pressing is the act of applying pressure to grapes or pomace in order to separate juice or wine from grapes and grape skins. Pressing is not always a necessary act in winemaking; if grapes are crushed there is a considerable amount of juice immediately liberated (called free-run juice) that can be used for vinification. Typically this free-run juice is of a higher quality than the press juice. However, most wineries do use presses in order to increase their production (gallons) per ton, as pressed juice can represent between 15%-30% of the total juice volume from the grape. Presses act by positioning the grape skins or whole grape clusters between a rigid surface and a moveable surface and slowly decrease the volume between the two surfaces. Modern presses dictate the duration and pressure at each press cycle, usually ramping from 0 Bar to 2.0 Bar. Sometimes winemakers choose pressures which separate the streams of pressed juice, called making "press cuts." As the pressure increases the amount of tannin extracted from the skins into the juice increases, often rendering the pressed juice excessively tannic or harsh. Because of the location of grape juice constituents in the berry (water and acid are found primarily in the mesocarp or pulp, whereas tannins are found primarily in the pericarp, or skin, and seeds), pressed juice or wine tends to be lower in acidity with a higher pH than the free-run juice.

Before the advent of modern winemaking, most presses were basket presses made of wood and operated manually. Basket presses are composed of a cylinder of wooden slats on top of a fixed plate, with a moveable plate that can be forced downward (usually by a central ratcheting threaded screw.) The press operator would load the grapes or pomace into the wooden cylinder, put the top plate in place and lower it until juice flowed from the wooden slats. As the juice flow decreased, the plate was ratcheted down again. This process continued until the press operator determined that the quality of the pressed juice or wine was below standard, or all liquids had been pressed. Since the early 1990s, modern mechanical basket presses have been revived through higher-end producers seeking to replicate the gentle pressing of the historical basket presses. Because basket presses have a relatively compact design, the press cake offers a relatively longer pathway for the juice to travel before leaving the press. It is believed by advocates of basket presses that this relatively long pathway through the grape or pomace cake serves as a filter to solids that would otherwise affect the quality of the press juice. With red wines, the must is pressed after primary fermentation, which separates the skins and other solid matter from the liquid. With white wine, the liquid is separated from the must before fermentation. With rose, the skins may be kept in contact for a shorter period to give colour to the wine, in that case the must may be pressed as well. After a period in which the wine stands or ages, the wine is separated from the dead yeast and any solids that remained (called lees), and transferred to a new container where any additional fermentation may take place.

Pigeage

Pigeage is a French winemaking term for the traditional stomping of grapes in open fermentation tanks. To make certain types of wine, grapes are put through a crusher and then poured into open fermentation tanks. Once fermentation begins, the grape skins are pushed to the surface by carbon dioxide gases released in the fermentation process. This layer of skins and other solids is known as the cap. As the skins are the source of the tannins, the cap needs to be mixed through the liquid each day, or "punched," which traditionally is done by stomping through the vat.

Cold and Heat Stabilization

Cold stabilization is a process used in winemaking to reduce tartrate crystals (generally potassium bitartrate) in wine. These tartrate crystals look like grains of clear sand, and are also known as "wine crystals" or "wine diamonds". They are formed by the union of tartaric acid and potassium, and may appear to be sediment in the wine, though they are not. During the cold stabilizing process after fermentation, the temperature of the wine is dropped to close to freezing for 1–2 weeks. This will cause the crystals to separate from the wine and stick to the sides of the holding vessel. When the wine is drained from the vessels, the tartrates are left behind. They may also form in wine bottles that have been stored under very cold conditions. During "heat stabilization", unstable proteins are removed by adsorption onto bentonite, preventing them from precipitating in the bottled wine.

Secondary Fermentation and Bulk Aging

During the secondary fermentation and aging process, which takes three to six months, the fermentation continues very slowly. The wine is kept under an airlock to protect the wine from oxidation. Proteins from the grape are broken down and the remaining yeast cells and other fine particles from the grapes are allowed to settle. Potassium bitartrate will also precipitate, a process which can be enhanced by cold stabilization to prevent the appearance of (harmless) tartrate crystals after bottling. The result of these processes is that the originally cloudy wine becomes clear. The wine can be racked during this process to remove the lees.

The secondary fermentation usually takes place in either large stainless steel vessels with a volume of several cubic meters, or oak barrels, depending on the goals of the winemakers. Unoaked wine is fermented in a barrel made of stainless steel or other material having no influence in the final taste of the wine. Depending on the desired taste, it could be fermented mainly in stainless steel to be briefly put in oak, or have the complete fermentation done in stainless steel. Oak could be added as chips used with a non-wooden barrel instead of a fully wooden barrel. This process is mainly used in cheaper wine.

Amateur winemakers often use glass carboys in the production of their wine; these vessels (sometimes called *demijohns*) have a capacity of 4.5 to 54 litres (1.2–14.3 US gallons). The kind of vessel used depends on the amount of wine that is being made, the grapes being used, and the intentions of the winemaker.

Malolactic Fermentation

Malolactic fermentation occurs when lactic acid bacteria metabolize malic acid and produce lactic acid and carbon dioxide. This is carried out either as an intentional procedure in which specially cultivated strains of such bacteria are introduced into the maturing wine, or it can happen by chance if uncultivated lactic acid bacteria are present.

Malolactic fermentation can improve the taste of wine that has high levels of malic acid, because malic acid in higher concentration generally causes an often unpleasant harsh and bitter taste sensation, whereas lactic acid is perceived as more gentle and less sour.

The process is used in most red wines and is discretionary for white wines.

Laboratory Tests

Whether the wine is aging in tanks or barrels, tests are run periodically in a laboratory to check the status of the wine. Common tests include °Brix, pH, titratable acidity, residual sugar, free or available sulfur, total sulfur, volatile acidity and percent alcohol. Additional tests include those for the crystallization of cream of tartar (potassium hydrogen tratrate) and the precipation of heat unstable protein; this last test is limited to white wines. These tests are often performed throughout the making of the wine as well as prior to bottling. In response to the results of these tests, a winemaker can then decide on appropriate remedial action, for example the addition of more sulfur dioxide. Sensory tests will also be performed and again in response to these a wine maker may take remedial action such as the addition of a protein to soften the taste of the wine.

Brix is one measure of the soluble solids in the grape juice and represents not only the sugars but also includes many other soluble substances such as salts, acids and tannins, sometimes called Total Soluble Solids (TSS). However, sugar is by far the compound in greatest quantity and so for all practical purposes these units are a measure of sugar level. The level of sugar in the grapes is important not only because it will determine the final alcohol content of the wine, but also because it is an indirect index of grape maturity. Brix (Bx for short) is measured in grams per hundred grams of solution, so 20 Bx means that 100 grams of juice contains 20gm of dissolved compounds. There are other common measures of sugar content of grapes, Specific gravity, Oechsle (Germany) and Beaume (France). The French Baume has the benefit that one Be° gives approximately one percent alcohol. Also one Be° is equal to 1.8 Brix, that is 1.8 grams of sugar per one hundred grams. This helps with deciding how much sugar to add if the juice is low in sugar: to achieve one percent alcohol add 1.8 grams per 100 ml or 18 grams per liter. This process is called chaptalization and is illegal in some countries (but perfectly acceptable for the home winemaker.) Generally, for the making of dry table wines a Bx of between 20 and 25 is desirable (equivalent to Be° of 11 to 14.)

A Brix test can be run either in the lab or in the field for a quick reference number to see what the sugar content is. Brix is usually measured with a refractometer while the other methods use a hydrometer. Generally, hydrometers are a cheaper alternative. For more accurate use of sugar measurement it should be remembered that all measurements are affected by the temperature at which the reading is made. Suppliers of equipment generally will supply correction charts.

Volatile acidity test verifies if there is any steam distillable acids in the wine. Mainly present is acetic acid but lactic, butyric, propionic and formic acids can also be found. Usually the test checks for these acids in a cash still, but there are new methods available such as HPLC, gas chromatography and enzymatic methods. The amount of volatile acidity found in sound grapes is negligible, since it is a by-product of microbial metabolism. It's important to remember that acetic acid bacteria require oxygen to grow. Eliminating any air in wine containers as well as a sulfur dioxide addition will limit their growth. Rejecting moldy grapes will also prevent possible problems associated with acetic acid bacteria. Use of sulfur dioxide and inoculation with a low-V.A. producing strain of Saccharomyces may deter acetic acid producing yeast. A relatively new method for removal of volatile acidity from a wine is reverse osmosis. Blending may also help—a wine with high V.A. can be filtered (to remove the microbe responsible) and blended with a low V.A. wine, so that the acetic acid level is below the sensory threshold.

Blending and Fining

Different batches of wine can be mixed before bottling in order to achieve the desired taste. The winemaker can correct perceived inadequacies by mixing wines from different grapes and batches that were produced under different conditions. These adjustments can be as simple as adjusting acid or tannin levels, to as complex as blending different varieties or vintages to achieve a consistent taste.

Fining agents are used during winemaking to remove tannins, reduce astringency and remove microscopic particles that could cloud the wines. The winemakers decide on which fining agents are used and these may vary from product to product and even batch to batch (usually depending on the grapes of that particular year). Gelatin has been used in winemaking for centuries and is recognized as a traditional method for wine fining, or clarifying. It is also the most commonly used agent to reduce the tannin content. Generally no gelatin remains in the wine because it reacts with the wine components, as it clarifies, and forms a sediment which is removed by filtration prior to bottling.

Besides gelatin, other fining agents for wine are often derived from animal and fish products, such as micronized potassium casseinate (casein is milk protein), egg whites, egg albumin, bone char, bull's blood, isinglass (Sturgeon bladder), PVPP (a synthetic compound), lysozyme, and skim milk powder.

Some aromatized wines contain honey or egg-yolk extract.

Non-animal-based filtering agents are also often used, such as bentonite (a volcanic clay-based filter), diatomaceous earth, cellulose pads, paper filters and membrane filters (thin films of plastic polymer material having uniformly sized holes).

Fermentation-Turning Grape Juice into Alcohol

Grape juice is turned into alcohol by the process of "fermentation." Grapes on the vine are covered with yeast, mold and bacteria. By putting grape juice into a container at the right temperature, yeast will turn the sugar in the juice into alcohol and carbon dioxide. The grape juice will have fermented. Fermentation is carried out in stainless steel vessels.

Yeast also gives flavor to the wine. But the yeast that is on the grape skin when it is harvested may not have the desired flavor. Other things on the outside of a grape are not good for wine (for example, acetic bacteria on the grapes can cause the wine to turn to vinegar). The winemaker can eliminate unwanted yeast's, molds and bacteria, most commonly by using the "universal disinfectant," sulfur dioxide. Unfortunately, the sulfites which remain in the wine may cause a lot of discomfort to some wine drinkers. Some winemakers prefer NOT to do this, and purposely create wines that are subject to the vagaries (and different flavours) of the yeast that pre-exist on the grapes ("wild yeast fermentation").

The winemaker has many different yeast strains to choose from (and can use different strains at different times during the process for better control fermentation). The most common wine yeast is Saccharomyces. This is a good point to stop and mention "Brett," also known as the Brettanomyces strain of yeast (which can be added or come from wild yeast fermentation). As yeast works, it causes grape juice ("must") to get hot. But if there's too much heat, the yeast won't work. Cooling coils are necessary to maintain a temperature below 45 C.

A less modern, but still wide widely used way to ferment wine is to place it in small oak barrels. "Barrel fermentation" is usually done at a lower temperature in temperature controlled rooms and takes longer, perhaps around 6 weeks. The longer fermentation and use of wood contributes to the flavour (and usually expense) of the wine.

The skins and pulp which remain in a red wine vat will rise to and float on top of the juice. This causes problems (if it dries out, it's a perfect breeding ground for injurious bacteria), so the winemaker will push this "cap" back down into the juice, usually at least twice a day. In large vats, this is accomplished by pumping juice from the bottom of the vat over the top of the cap.

Eventually the yeast is no longer changing sugar to alcohol (though different strains of yeast, which can survive in higher and higher levels of alcohol, can take over and contribute their own flavour to the wine-as well as converting a bit more sugar to alcohol).

After all this is completed what you have left is the wine, "dead" yeast cells, known as "lees and various other substances.

Malo-lactic Fermentation

The winemaker may choose to allow a white wine to undergo a second fermentation which occurs due to malic acid in the grape juice. When malic acid is allowed to break down into carbon dioxide and lactic acid (thanks to bacteria in the wine), it is known as "malo-lactic fermentation," which imparts additional flavour to the wine. A "buttery" flavour in some whites is due to this process. This process is used for sparkling wines.

First Racking

After fermentation completed naturally or stopped by addition of distilled spirit, first racking is carried out. This involves the wine to stand still until most yeast cells and fine suspended material settle out. The wine is then filtered without disturbing the sediment or the yeast.

Winery Aging

The winery may then keep the wine so that there can be additional clarification and, in some wines, to give it a more complex flavours. Flavour can come from wood (or more correctly from the chemicals that make up the wood and are taken up into the wine).

The wine may be barrel aged for several months to several years. No air is allowed to enter the barrels during this period.

Ignoring any additional processing that might be used, you could empty the barrels into bottles and sell your wine. However, during the winery aging, the smaller containers may develop differences. So the winemaker will probably "blend" wine from different barrels, to achieve a uniform result. Also, the winemaker may blend together different grape varieties to achieve desired characteristics.

Stabilization, Filtration

Stabilization is carried out to remove traces of tartaric acid. These tartarates present in the grape juice tend to crystallize in wine and if not removed completely can slowly reappear as glass like crystals in final bottles on storage.

Stabilization with respect to tartarates may involve chilling of wine that can crystallize tartarates and these crystals can be removed by filtration.

Pasteurization

If the wine has an alcohol content less than 14% it may be heat pasteurized or cold pasteurized through microporous filters just before bottling.

Preservatives

The most common preservative used in winemaking is sulfur dioxide, achieved by adding sodium or potassium metabisulphite. Another useful preservative is potassium sorbate.

Sulfur dioxide has two primary actions, firstly it is an anti microbial agent and secondly an anti oxidant. In the making of white wine it can be added prior to fermentation and immediately after alcoholic fermentation is complete. If added after alcoholic ferment it will have the effect of preventing or stopping malolactic fermentation, bacterial spoilage and help protect against the damaging effects of oxygen. Additions of up to 100 mg per liter (of sulfur dioxide) can be added, but the available or free sulfur dioxide should be measured by the aspiration method and adjusted to 30 mg per liter. Available sulfur dioxide should be maintained at this level until bottling. For rose wines smaller additions should be made and the available level should be no more than 30 mg per liter.

In the making of red wine sulfur dioxide may be used at high levels (100 mg per liter) prior to ferment to assist stabilize colour otherwise it is used at the end of malolactic ferment and performs the same functions as in white wine. However, small additions (say 20 mg per liter) should be used to avoid bleaching red pigments and the maintenance level should be about 20 mg per liter. Furthermore, small additions (say 20 mg per liter) may be made to red wine after alcoholic ferment and before malolactic ferment to overcome minor oxidation and prevent the growth of acetic acid bacteria.

Without the use of sulfur dioxide, wines can readily suffer bacterial spoilage no matter how hygienic the winemaking practice.

Potassium sorbate is effective for the control of fungal growth, including yeast, especially for sweet wines in bottle. However, one potential hazard is the metabolism of sorbate to geraniol a potent and very unpleasant by-product. To avoid this, either the wine must be sterile bottled or contain enough sulfur dioxide to inhibit the growth of bacteria. Sterile bottling includes the use of filtration.

Filtration

Filtration in winemaking is used to accomplish two objectives, clarification and microbial stabilization. In clarification, large particles that affect the visual appearance of the wine are removed. In microbial stabilization, organisms that affect the stability of the wine are removed therefore reducing the likelihood of re-fermentation or spoilage.

The process of clarification is concerned with the removal of particles; those larger than 5–10 micrometers for coarse polishing, particles larger than 1–4 micrometers for clarifying or polishing. Microbial stabilization requires a filtration of at least 0.65 micrometers. However, filtration at this level may lighten a wines colour and body. Microbial stabilization does not imply sterility. It simply means that a significant amount of yeast and bacteria have been removed.

Bottling Wine

Producers often use different shaped bottles to denote different types of wine. Coloured bottles help to reduce damage by light. (Light assists in oxidation and breakdown of the wine into chemicals, such as mercaptan, which are undesirable.)

Cellaring Wine

Most people assume that the longer that you keep a wine, the better it will get Since its best to store wine under certain conditions, like in a cool damp underground cellar, this is known as "cellaring" wine. It is a misconception that you MUST age wine. The fact is, throughout the world, most wine is drunk "young" (that is relatively soon after it is produced, perhaps 12 to 18 months), even wines that are "better" if aged. While some wines will "mature" and become better over time, others will not and should be drunk immediately, or within a few years. Tannin is a substance that comes from the seeds, stems and skins of grapes. Additional tannin can come from the wood during barrel aging in the winery. It is a preservative and is important to the long term maturing of wine.

Through time, tannin (which has a bitter flavour) will precipitate out of the wine (becoming sediment in the bottle) and the complexity of the wine's flavour from fruit, acid and all the myriad other substances that make up the wine's character will come into greater balance. Generally, it is red wines that are the ones that CAN (but do not have to be) produced with a fair amount of tannin with an eye towards long term storing and maturation.

The bad news is that you shouldn't drink it young since it will taste too harsh (and probably cost too much, besides). The good news is that after a number of years, what you get is a prized, complex and balanced wine.

Remember that red wines get their colour from the stems and skins of the grape. This gives the wine tannin and aging capacity. White wines may have no contact with the stems and skins and will have little tannin (though some can be added, again, through barrel aging). Therefore most white wines don't age well. Even the ones which do get better through time will not last nearly as long as their red cousins. A fair average for many "ageable" whites would be about 5 to 7 years (some might go 10). On the other hand, really "ageable" reds can easily be kept for 30 years and longer.

Storing Wine

For wines that should be aged, a cellar should have proper :Temperature which does not have rapid fluctuation. 55 degrees Fahrenheit is a good, but you can live with 50 to 57 degrees Fahrenheit (10 to 14 degrees Centigrade). Wide swings in temperature will harm the wine. Having too high a temperature will age the wine faster so it won't get as complex as it might have. Having too low a temperature will slow the wine's maturation. Humidity. About 60 percent is right. This helps keep the cork moist. The wine will oxidize if the air (and its oxygen) gets to it. If the cork dries out, it can shrink and let air in. This is another reason to keep the bottles on their sides. The wine itself will help keep the cork moist.

Pectic Enzymes on Clarification of Wines

The enzymes such as pectinase that hydrolyze the large pectin molecules.

Pectinase

An enzyme that catalyses the hydrolysis of pectin molecules.

Pectin

A heavy, colloidal substance found in most ripe fruit which promotes the formation of gelatinous solutions and hazes in the finished wine. Fermenting fruit pulps with high pectin content, such as apples, should be treated with pectic enzyme, especially if the pulp is boiled to extract the fruit flavour (boiling releases the pectin, while pectic enzymes destroy it). Pectin is a structural heteropolysaccharide contained in the primary cell walls of terrestrial plants. It was first isolated and described in 1825 by Henri Braconnot. It is produced commercially as a white to light brown powder, mainly extracted from citrus fruits, and is used in food as a gelling agent particularly in jams and jellies. It is also used in fillings, sweets, as a stabilizer in fruit juices and milk drinks and as a source of dietary fibre.

Biology

In plant cells, pectin consists of a complex set of polysaccharides that are present in most primary cell walls and particularly abundant in the non-woody parts of terrestrial plants. Pectin is present not only throughout primary cell walls but also in the middle lamella between plant cells where it helps to bind cells together.

The amount, structure and chemical composition of pectin differs between plants, within a plant over time and in different parts of a plant. During ripening, pectin is broken down by the enzymes pectinase and pectinesterase; in this process the fruit becomes softer as the middle lamella breaks down and cells become separated from each other. A similar process of cell separation caused by pectin breakdown occurs in the abscission zone of the petioles of deciduous plants at leaf fall.

Pectin is a natural part of human diet, but does not contribute significantly to nutrition. The daily intake of pectin from fruits and vegetables can be estimated to be around 5 g (assuming consumption of approximately 500 g fruits and vegetables per day).

In human digestion, pectin goes through the small intestine more or less intact. Pectin is thus a soluble dietary fibre.

Consumption of pectin has been shown to reduce blood cholesterol levels. The mechanism appears to be an increase of viscosity in the intestinal tract, leading to a reduced absorption of cholesterol from bile or food. In the large intestine and colon, microorganisms degrade pectin and liberate short-chain fatty acids that have positive influence on health (prebiotic effect).

Chemistry

Pectins are a family of complex polysaccharides that contain 1,4-linked α-D-galactosyluronic acid residues. Three pectic polysaccharides have been isolated from plant primary cell walls and structurally characterized. These are:

- Homogalacturonans
- Substituted galacturonans
- Rhamnogalacturonans.

Homogalacturonans are linear chains of α-(1-4)-linked D-galacturonic acid.

Substituted galacturonans are characterized by the presence of saccharide appendant residues (such as D-xylose or D-apiose in the respective cases of xylogalacturonan and apiogalacturonan) branching from a backbone of D-galacturonic acid residues.

Rhamnogalacturonan I pectins (RG-I) contain a backbone of the repeating disaccharide: 4)-α-D-galacturonic acid-(1,2)-α-L-rhamnose-(1. From many of the rhamnose residues, sidechains of various neutral sugars branch off. The neutral sugars are mainly D-galactose, L-arabinose and D-xylose, the types and proportions of neutral sugars varying with the origin of pectin. Another structural type of pectin is rhamnogalacturonan II (RG-II), which is a less frequent complex, highly branched polysaccharide. Rhamnogalacturonan II is classified by some authors within the group of substituted galacturonans since the rhamnogalacturonan II backbone is made exclusively of D-galacturonic acid units.

Isolated pectin has a molecular weight of typically 60–130,000 g/mol, varying with origin and extraction conditions.

In nature, around 80% of carboxyl groups of galacturonic acid are esterified with methanol. This proportion is decreased more or less during pectin extraction. The ratio of esterified to non-esterified galacturonic acid determines the Behaviour of pectin in food applications. This is why pectins are classified as high- vs. low-ester pectins – or in short HM vs. LM-pectins, with more or less than half of all the galacturonic acid esterified.

The non-esterified galacturonic acid units can be either free acids (carboxyl groups) or salts with sodium, potassium or calcium. The salts of partially esterified pectins are called pectinates, if the degree of esterification is below 5% the salts are called pectates, the insoluble acid form, pectic acid.

Some plants like sugar beet, potatoes and pears contain pectins with acetylated galacturonic acid in addition to methyl esters. Acetylation prevents gel-formation but increases the stabilising and emulsifying effects of pectin. Amidated pectin is a modified form of pectin. Here, some of the galacturonic acid is converted with ammonia to carboxylic acid amide. These pectins are more tolerant of varying calcium concentrations that occur in use. To prepare a pectin-gel, the ingredients are heated, dissolving the pectin. Upon cooling below gelling temperature, a gel starts to form. If gel formation is too strong, syneresis or a granular texture are the result, whilst weak gelling leads to excessively soft gels. In high-ester pectins at soluble solids content above 60% and a pH-value between 2.8 and 3.6, hydrogen bonds and hydrophobic interactions bind the individual pectin chains together. These bonds form as water is bound by sugar and forces pectin strands to stick together. These form a 3-dimensional molecular net that creates the macromolecular gel. The gelling-mechanism is called a low-water-activity gel or sugar-acid-pectin gel.

In low-ester pectins, ionic bridges are formed between calcium ions and the ionised carboxyl groups of the galacturonic acid. This is idealised in the so-called "egg box-model". Low-ester pectins need calcium to form a gel, but can do so at lower soluble solids and higher pH-values than high-ester pectins. Amidated pectins behave like low-ester pectins but need less calcium and are more tolerant of excess calcium. Also, gels from amidated pectin are thermo-reversible – they can be heated and after cooling solidify again, whereas conventional pectin-gels will afterwards remain liquid.

High-ester pectins set at higher temperatures than low-ester pectins. However, gelling reactions with calcium increase as the degree of esterification falls. Similarly, lower pH-values or higher soluble solids (normally sugars) increase gelling speed. Suitable pectins can therefore be selected for jams and for jellies, or for higher sugar confectionery jellies.

Sources and Production

Apples, guavas, quince, plums, gooseberries, oranges and other citrus fruits, contain large amounts of pectin, while soft fruits like cherries, grapes and strawberries contain small amounts of pectin.

Typical levels of pectin in plants are (fresh weight):

- apples, 1–1.5%
- apricot, 1%
- cherries, 0.4%
- oranges 0.5–3.5%
- carrots approx. 1.4%
- citrus peels, 30%.

The main raw-materials for pectin production are dried citrus peel or apple pomace, both by-products of juice production. Pomace from sugar-beet is also used to a small extent.

From these materials, pectin is extracted by adding hot dilute acid at pH-values from 1.5 – 3.5. During several hours of extraction, the protopectin loses some of its branching and chain-length and goes into solution. After filtering, the extract is concentrated in vacuum and the pectin then precipitated by adding ethanol or isopropanol. An old technique of precipitating pectin with aluminium salts is no longer used (apart from alcohols and polyvalent cations; pectin also precipitates with proteins and detergents).

Alcohol-precipitated pectin is then separated, washed and dried. Treating the initial pectin with dilute acid leads to low-esterified pectins. When this process includes ammonium hydroxide, amidated pectins are obtained. After drying and milling pectin is usually standardised with sugar and sometimes calcium-salts or organic acids to have optimum performance in a particular application.

Worldwide, approximately 40,000 metric tons of pectin are produced every year.

Uses

The main use for pectin is as a gelling agent, thickening agent and stabilizer in food. The classical application is giving the jelly-like consistency to jams or marmalades, which would otherwise be sweet juices. For household use, pectin is an ingredient in gelling sugar (also known as "Jam Sugar") where it is diluted to the right concentration with sugar and some citric acid to adjust pH. In some countries, pectin is also available as a solution or an extract, or as a blended powder, for home jam making. For conventional jams and marmalades that contain above 60% sugar and soluble fruit solids, high-ester pectins are used. With low-ester pectins and amidated pectins less sugar is needed, so that diet products can be made. Pectin can also be used to stabilize acidic protein drinks, such as drinking yogurt, and as a fat substitute in baked goods. Typical levels of pectin used as a food additive are between 0.5 – 1.0% - this is about the same amount of pectin as in fresh fruit.

In medicine, pectin increases viscosity and volume of stool so that it is used against constipation and diarrhea. Until 2002, it was one of the main ingredients used in Kaopectate, along with kaolinite. Pectin is also used in throat lozenges as a demulcent. In cosmetic products, pectin acts as stabilizer. Pectin is also used in wound healing preparations and speciality medical adhesives, such as colostomy devices. Also, it is considered a natural remedy for nausea. Pectin rich foods are proven to help nausea.

In ruminant nutrition, depending on the extent of lignification of the cell wall, pectin is up to 90% digestible by bacterial enzymes. Ruminant nutritionists recommend that the digestibility and energy concentration in forages can be improved by increasing pectin concentration in the forage.

In the cigar industry, pectin is considered an excellent substitute for vegetable glue and many cigar smokers and collectors will use pectin for repairing damaged tobacco wrapper leaves on their cigars. Pectin is also used in jellybeans.

The Effect of Pectic Enzymes in Wine Making

What are Pectins? How do they Influence the Clarification of Wine?

Broadly speaking, pectins constitute a group of closely related carbohydrates allied to such polysaccharides as starch and chemically defined as complex, methylated polygalacturonic acids. They possess colloidal properties and will gel (i.e. set to jelly in the presence of high concentrations of sugar and acid). The essential role of pectins as setting agents in jams and jellies has long been recognized. For making jam and jelly fruits which are naturally high in pectin are a decided advantage. Exactly the opposite is true in winemaking where even a small amount of pectin can cause many problems.

Although winemakers rarely encounter conditions which cause musts and wines high in pectin to gel, the fact that some is present can have other equally undesirable effects. Thus, wines containing pectin are exceptionally difficult to clear because the pectin acts as a colloid and stabilizes the haze. The minute particles which are responsible for the wine appearing cloudy are too small to settle out naturally no matter how long a period of storage is allowed. Wines only clear because the tiny suspended particles constituting the haze are mutually attracted and combined to form larger particles whose size eventually increases sufficiently for them to begin settling under the force of gravity. The action of protective colloid such as pectin and starch is to oppose these processes of clarification by stabilizing the haze forming particles and preventing their agglomeration. Since there is no tendency for the particles to attract each other, there is no particle growth and hence no settling so the wine remains stubbornly cloudy indefinitely. Pectin stabilizes haze so efficiently that neither fining nor filtering is likely to be effective and may only make matters worse. How, then, can wines whose haze forming particles are stabilized by pectin be clarified? The only satisfactory way to clear pectin clouded wines is to degrade the pectin so that the protective action on the particles causing the haze is destroyed and agglomeration can occur in the normal manner. This is accomplished by the addition of ½ teaspoon of Pectic Enzyme Powder per gallon of wine. The essential function of these enzymes is to break the long pectin chain into shorter units which are unable to act as protective colloids.

Although the addition of Pectic Enzymes can be used with great success for clearing wines with pectin stabilized hazes, it is obviously preferable to avoid encountering such difficulties at such a late stage in the production of the wine. Far less trouble would occur if the pectin could be destroyed prior to and/or during fermentation so that it is completely degraded by the time the first rack occurs. The wine would then clear more quickly and more completely. So for best results, add ½ teaspoon of Pectic Enzyme Powder to all your wines prior to the addition of yeast.

When fruit is added to beer or mead, it should also be treated with Pectic Enzymes to prevent pectin stabilized hazes.

1. Yields of free run and total yields of wine per ton of grapes have on the average been greater from enzyme treated than from untreated crushed grapes.
2. Use of pectic enzymes in crushed white grapes almost invariably deepened the colour of the wine in comparison with that of wine made from grapes to which no enzyme was added—probably owing to greater extraction of colour from the skins.
3. Colour of wines made from free run juice to which was added a pectic enzyme was in mamy cases about like that of the untreated wine, but also in about an equal number of cases the colour of the treated wines on aging became slightly deeper (yellower) than that of the untreated wines.

4. Therefore, if light colour is the primary objective, the enzyme should be added to the white juice rather than to the crushed grapes.
5. Enzyme-treated wines have invariably matured more rapidly than the untreated. The wine maker must be careful not to let the treated wines overmature before bottling. In one winery experiment of the 1950 season the enzyme-treated wine is now inferior to the untreated in flavour and bouquet. In other cases there is little difference, or the enzyme treated is superior.
6. Enzyme-treated wines developed much more compact lees than the treated and were racked with less danger of carrying lees over into the wine.
7. As in previous seasons (1936-1949), we found that the enzyme-treated wines cleared much more rapidly than the untreated.
8. Also, they filtered much more rapidly and much more wine could be filtered with each setting of the filter.
9. Much less bentonite was needed for clarification of the enzyme-treated wines—usually 1-2 pounds per 1000 gallons as against 4-10 pounds for the untreated.
10. At two winners relatively small amounts of Pectineal have been used to pound of Pectinol-0 per ton. The wines so treated do not clear as rapidly as with larger additions of enzyme but are found to filter very much more rapidly than the untreated and to require far less bentonite in clarification.
11. Enzyme-treated juices were found commercially to filter more readily and to give concentrates of better quality than the untreated.
12. As reported by Berg and Marsh, it was observed that Pectineal greatly increased the rate of clearing of heat extracted fermented red musts. Usually the enzyme-treated heat extracted wines in our experiments were of deeper colour than the checks, indicating that the enzyme may have stabilized the colour.
13. On the other hand, where the enzyme was used in the crushed grapes without heat extraction the colour of the resulting dry wines were in some cases less intense than in the check lots. In others it was more intense or about equal to the untreated.
14. No very clear cut differences were observed in the effect of several different Rohm and Haas enzyme preparations on the colour of either red or white wines.

15. Enzyme-treated white wines darkened more rapidly than the untreated when incubated at 45° C (131° F), although the difference was usually not very great.
16. Preliminary evidence was found to indicate that galacturonic acid formed by hydrolysis of the pectin of the must by the enzyme may be concerned in the darkening of the colour of enzyme-treated wines.
17. Pectineal solutions of the concentrations applied in wine making did not darken; indicating that where darkening of the treated wine occurs it is some constituent of the wine itself that is concerned.
18. Ports made with Pectineal were in most cases deeper in colour than the untreated.
19. Enzyme-treated musts foamed much less than the untreated during fermentationo
20. There appeared to be no consistent differences in heat and cold stability of enzyme-treated and untreated wines. Finished enzyme-treated dry white wines were stable at 131° F and at 0° F.
21. Maintenance of a reasonable level of SO_2 in enzyme-treated white wines held the colour.

Vinegar Fermentation and Production

Vinegar is an acid liquid produced from the fermentation of ethanol in a process that yields its key ingredient, acetic acid (ethanoic acid). It also may come in a diluted form. The acetic acid concentration typically ranges from 4% to 8% by volume for table vinegar and up to 18% for pickling. Natural vinegars also contain small amounts of tartaric acid, citric acid, and other acids. Vinegar has been used since ancient times and is an important element in European, Asian, and other cuisines.

The word "vinegar" derives from the Old French *vin aigre*, meaning "sour wine", which in turn is derived from the Latin "vinum aegrum" meaning "feeble wine".

History

Vinegar has been made and used for thousands of years. Traces of it have been found in Egyptian urns dating from around 3000 BC. According to Shennong's *Herb Classic*, vinegar was invented in China during the Xia Dynasty, around 2000 BC.

In the Tanakh and the Old Testament of the Bible, it is mentioned as unpleasant to drink (Ps. 69:21) and foolish to combine with *nether* (most likely soda ash, although possibly potash, natron, or niter) (Prov. 25:20), but more favorably as a condiment when Boaz allows Ruth to "dip her piece of bread in the vinegar" (Ruth 2:14). Jesus was offered vinegar or sour wine while on the cross (Matthew 27:48; Mark 15:36). In Islamic traditions, vinegar is one of the four favoured condiments of the Prophet Muhammad, who called it a "Blessed seasoning".

In 1864, Louis Pasteur showed that vinegar results from a natural fermentation process.

Production

Vinegar is made from the fermentation of ethanol by acetic acid bacteria. The ethanol may be derived from many different sources including wine, cider, beer or fermented fruit juice, or it may be made synthetically from natural gas and petroleum derivatives.

Commercial vinegar is produced either by fast or slow fermentation processes. Slow methods generally are used with traditional vinegars and fermentation proceeds slowly over the course of weeks or months.

The longer fermentation period allows for the accumulation of a nontoxic slime composed of acetic acid bacteria and soluble cellulose, known as the mother of vinegar. Fast methods add mother of vinegar (i.e., bacterial culture) to the source liquid before adding air using a Venturi pump system or a turbine to promote oxygenation to obtain the fastest fermentation. In fast production processes, vinegar may be produced in a period ranging from 20 hours to three days.

Varieties

Malt

Malt vinegar is made by malting barley, causing the starch in the grain to turn to maltose. Then an ale is brewed from the maltose and allowed to turn into vinegar, which is then aged. It typically is light brown in colour.

Wine

Wine vinegar is made from red or white wine and is the most commonly used vinegar in Mediterranean countries and Central Europe. As with wine, there is a considerable range in quality. Better quality wine vinegars are matured in wood for up to two years and exhibit a complex, mellow flavour. Wine vinegar tends to have a lower acidity than that of white or cider vinegars. There are more expensive wine vinegars that are made from individual varieties of wine, such as Champagne, Sherry, or pinot grigio.

Apple Cider

Apple cider vinegar, otherwise known simply as cider vinegar or ACV, is made from cider or apple must and has a brownish-yellow colour. It often is sold unfiltered and unpasteurized with the mother of vinegar present, as a natural product. It is very popular, partly because of supposed beneficial health and beauty properties. Because of its acidity, apple cider vinegar may be very harsh, even burning, to the throat. If taken straight, (as opposed to used in cooking), it can be diluted (e.g., with fruit juice or water) before drinking. It is also sometimes sweetened with sugar or honey. There have been reports of acid chemical burns of the throat from apple cider vinegar tablets, but "doubt remains as to whether apple cider vinegar was in fact an ingredient in the evaluated products."

Fruit

Fruit vinegars are made from fruit wines, usually without any additional flavouring. Common flavours of fruit vinegar include apple, black currant, raspberry, quince, and tomato. Typically, the flavours of the original fruits remain in the final product.

Most fruit vinegars are produced in Europe, where there is a growing market for high-priced vinegars made solely from specific fruits (as opposed to non-fruit vinegars which are infused with fruits or fruit flavours). Several varieties, however, also are produced in Asia. Persimmon vinegar, called *gam sikcho*, is popular in South Korea. Jujube vinegar and wolfberry vinegar are produced in China.

Umezu, a salty, sour liquid that is a by-product of *umeboshi* (pickled *ume*) production, is produced in Japan, but technically is not a true vinegar.

Jamun Sirka is a vinegar produced from the *Jamun* (or rose apple) fruit in India. It is considered to be medicinally valuable for stomach, spleen and diabetic ailments.

Balsamic

Balsamic vinegar is an aromatic, aged type of vinegar traditionally crafted in the Modena and Reggio Emilia provinces of Italy from the concentrated juice, or must, of white grapes (typically of the Trebbiano variety). It is very dark brown in colour and its flavour is rich, sweet, and complex, with the finest grades being the product of years of aging in a successive number of casks made of various types of wood (including oak, mulberry, chestnut, cherry, juniper, ash, and acacia). Originally a product available only to the Italian upper classes, a cheaper form of balsamic vinegar became widely known and available around the world in the late twentieth century.

True balsamic vinegar (which has Protected Designation of Origin) is aged for 12 to 25 years. Balsamic vinegars that have been aged for up to 100 years are available, though they are usually very expensive. The commercial balsamic sold in supermarkets is typically made with concentrated grape juice mixed with a strong vinegar, which is laced with caramel and sugar. Regardless of how it is produced, balsamic vinegar must be made from a grape product. Balsamic vinegar has a high acidity level but the tart flavour is usually hidden by the sweetness of the other ingredients, making it very mellow.

Rice

Rice vinegar is most popular in the cuisines of East and Southeast Asia. It is available in "white" (light yellow), red, and black varieties. The Japanese prefer a light rice vinegar for the preparation of sushi rice and salad dressings. Red rice vinegar traditionally is coloured with red yeast rice. Black rice vinegar (made with black glutinous rice) is most popular in China, and it is also widely used in other east Asian countries. White rice vinegar has a mild acidity and a somewhat "flat", uncomplex flavour. Some varieties of rice vinegar are sweetened or otherwise seasoned with spices or other added flavorings.

Coconut

Coconut vinegar, made from fermented coconut water, is used extensively in Southeast Asian cuisine (particularly in the Philippines, a major producer, where it is called *suka ng niyog*), as well as in some cuisines of India. A cloudy white liquid, it has a particularly sharp, acidic taste with a slightly yeasty note.

Palm

Palm vinegar, made from the fermented sap from flower clusters of the nipa palm (also called attap palm), is used most often in the Philippines, where it is produced, and where it is called *sukang paombong*.

Cane

Cane vinegar, made from sugar cane juice, is most popular in the Philippines, in particular, the Ilocos Region of the northern Philippines (where it is called *sukang iloko*), although it also is produced in France and the United States. It ranges from dark yellow to golden brown in colour and has a mellow flavour, similar in some respects, to rice vinegar, though with a somewhat "fresher" taste. Contrary to expectation, containing no residual sugar, it is not sweeter than other vinegars. In the Philippines, it often is labelled as *suka na maasim*, although this is simply a generic term meaning "sour vinegar."

Cane vinegars from Ilocos also varies in two different types. The "basi" or the sweet vinegar and the "Suka" which is the sour one. The sweet vinegar is used as a wine in Ilokanos, while the other type of vinegar is used as a seasoning and preservative. A white variation has become quite popular in Brazil in recent years, where it is the cheapest type of vinegar sold. It is now common for other types of vinegar (made from wine, rice and apple cider) to be sold mixed with cane vinegar to lower the costs.

Raisin

Vinegar made from raisins, called *khal 'anab* is used in cuisines of the Middle East, and is produced therein. It is cloudy and medium brown in colour, with a mild flavour.

Date

Vinegar made from dates is a traditional product of the Middle East.

Beer

Vinegar made from beer is produced in the United Kingdom, Germany, Austria, and the Netherlands. Although its flavour depends on the particular type of beer from which it is made, it is often described as having a malty taste. That produced in Bavaria, is a light golden colour with a very sharp and not-overly-complex flavour.

Honey

Vinegar made from honey is rare, although commercially available honey vinegars are produced in Italy, France, Romania and Spain.

East Asian Black

Chinese black vinegar is an aged product made from rice, wheat, millet, sorghum, or a combination thereof. It has an inky black colour and a complex, malty flavour. There is no fixed recipe and thus some Chinese black vinegars may contain added sugar, spices, or caramel colour. The most popular variety, Zhenjiang vinegar, originated in the city of Zhenjiang, in the eastern coastal province of Jiangsu, China and also is produced in Tianjin and Hong Kong.

A somewhat lighter form of black vinegar, made from rice, also is produced in Japan, where it is called *kurozu*. Since 2004 it has been marketed as a healthful drink; its manufacturers claim that it contains high concentrations of amino acids.

Flavored Vinegars

Popular fruit-flavoured vinegars include those infused with whole raspberries, blueberries, or figs (or else from flavourings derived from these fruits). Some of the more exotic fruit-flavoured vinegars include blood orange and pear.

Herb vinegars are flavoured with herbs, most commonly Mediterranean herbs such as thyme or oregano. Such vinegars can be prepared at home by adding sprigs of fresh or dried herbs to vinegar purchased at a grocery store; generally a light-coloured, mild tasting vinegar, such as that made from white wine, is used for this purpose.

Sweetened vinegar is of Cantonese origin and is made from rice wine, sugar and herbs including ginger, cloves, and other spices.

Job's Tears

In Japan, an aged vinegar also is made from Job's Tears, a tall grain-bearing tropical plant. The vinegar is similar in flavour to rice vinegar.

Kombucha

Kombucha vinegar is made from kombucha, a symbiotic culture of yeast and bacteria. The bacteria produce a complex array of nutrients and populate the vinegar with bacteria which some claim promotes a healthy digestive tract, although no scientific studies have confirmed this. Kombucha vinegar primarily is used to make a vinaigrette and is flavoured by adding strawberries, blackberries, mint, or blueberries at the beginning of fermentation.

Kiwifruit

A by-product of commercial kiwifruit growing is a large amount of waste in the form of firstly misshapen or otherwise rejected fruit that may constitute up to 30 per cent of the crop and secondly kiwifruit pomace which is the presscake residue left after kiwifruit juice manufacture. One of the uses for this waste is the production of kiwifruit vinegar. Produced commercially in New Zealand since, at least, the early 1990s Production of kiwifruit vinegar began for domestic sale in China in 2008.

Sinamak

A variation of cane vinegar from the Philippines (*sukang maasim*) is called *sinamak* which is simply a spiced version that mixes the cane vinegar with siling labuyo, onions and garlic.

Distilled Vinegar

Any type of vinegar may be distilled to produce a colourless solution of about 5% to 8% acetic acid in water. This is variously known as distilled spirit or "virgin vinegar", or white vinegar, and is used for medicinal, laboratory and cleaning purposes as well as in cooking, baking, meat preserving, including pickling.

The most common starting material, because of its low cost, is malt vinegar. By FDA regulations since the 1950s, Distilled Vinegar may be synthetic ethanol made by direct chemical oxidation of wood or fossil fuels. Distilled and other highly processed vinegars are mineral deficient and, when consumed, pull calcium and other minerals from the bones and tissues.

Spirit Vinegar

The term 'spirit vinegar' is sometimes reserved for the stronger variety (5% to 20% acetic acid) made from sugar cane or from chemically produced acetic acid.

Culinary Uses

Vinegar is commonly used in food preparation, particularly in pickling processes, vinaigrettes, and other salad dressings. It is an ingredient in sauces such as mustard, ketchup, and mayonnaise. Vinegar is sometimes used while making chutneys. It is often used as a condiment. Marinades often contain vinegar.

- Condiment for beetroot — cold, cooked beetroot is commonly eaten with vinegar
- Condiment for fish and chips — People commonly use malt vinegar (or non-brewed condiment) on chips.
- Flavouring for potato chips — many American, Canadian and British manufacturers of packaged potato chips and crisps feature a variety flavoured with vinegar and salt.
- Vinegar pie — a North American dessert made with a vinegar to one's taste and similar to chess pie.
- Pickling — any vinegar can be used to pickle foods.
- Cider vinegar and sauces — cider vinegar usually is not suitable for use in delicate sauces.

- Substitute for fresh lemon juice — cider vinegar can usually be substituted for fresh lemon juice in recipes and obtain a pleasing effect although it lacks the vitamin C.
- Saucing roast lamb — pouring cider vinegar over the meat when roasting lamb, especially when combined with honey or when sliced onions have been added to the roasting pan, produces a sauce.
- Sweetened vinegar is used in the dish of pork knuckles and ginger stew which is made among Chinese people of Cantonese backgrounds to celebrate the arrival of a new child.
- Sushi rice — Japanese use rice vinegar as an essential ingredient for sushi rice.
- Red vinegar — Sometimes used in Chinese soups
- Flavouring — used in the Southern U.S. to flavour collard greens, green beans, black-eyed peas, or cabbage to taste.
- Commonly put into mint sauce, for general palate preference.
- Vinegar—especially the coconut, cane, or palm variety—is one of the principal ingredients of Philippine.

Medical Uses

Many remedies and treatments have been ascribed to vinegar over millennia and in many different cultures, however, few have been verifiable using controlled medical trials and many that are effective to some degree have significant side effects and carry the possibility of serious health risks.

Soothing for Sunburns

White vinegar applied as a spray to tissue draped over a sunburn helps restore the lost acidic level to the skin, and gives a cooling effect.

Possible Cholesterol and Triacylglycerol Effects

A 2006 study concluded that a test group of rats fed with acetic acid (the main component of vinegar) had "significantly lower values for serum total cholesterol and triacylglycerol", among other health benefits. Rats fed vinegar or acetic acid have lower blood pressure than controls, although the effect has not been tested in humans. Reduced risk of fatal ischemic heart disease was observed among participants in a trial who ate vinegar and oil salad dressings frequently.

Blood Glucose Control and Diabetic Management

Prior to hypoglycemic agents, diabetics used vinegar teas to control their symptoms. Small amounts of vinegar (approximately 20 ml or two tablespoons of domestic vinegar) added to food, or taken along with a meal, have been shown by a number of medical trials to reduce the glycemic index of carbohydrate food for people with and without diabetes. This also has been expressed as lower glycemic index ratings in the region of 30%.

Diet Control

Multiple trials indicate that taking vinegar with food increases satiety (the feeling of fullness) and so, reduces the amount of food consumed.

Infections

Vinegar has been used to fight infections since Hippocrates, who lived between 460-377 BC, prescribed it for curing persistent coughs. As a result, vinegar is popularly believed to be effective against infections. While vinegar can be an effective antibacterial cleaning agent on hard surfaces such as washroom tiles and countertops, studies show that vinegar – whether taken internally or applied topically – is not effective against infections, lice, or warts.

Researchers at the Food Biotechnology Department, Instituto de la Grasa (CSIC) in Seville, Spain, conducted research on the antimicrobial activity of several products. Vinegar and red and white wines were among the products tested. (Note: The focus of the research was olive oil, but it confirmed other findings related to vinegar and red and white wines.) The following microorganisms were used in the study: *S. aureus, L. monocytogenes, S. Enteritidis, E.coli* 0157:H7, *S.sonnei* and Yersinia sp. Among the items tested, vinegar (5% acetic acid) showed the strongest bactericidal activity against all strains tested, which was attributed to its high acetic acid content.

The researchers noted their study confirmed previous results. It was noted that both red and white wines exhibited bactericidal activity, in particular against Salmonella Enteritidis and Yersinia sp. S. aureus and L. monocytogenes were the least sensitive to the wines.

Other Medicinal Uses

Applying vinegar to common jellyfish stings deactivates the nematocysts; however, placing the affected areas in hot water is a more effective treatment because the venom is deactivated by heat. The latter requires immersion in 45°C (113°F) water for at least four minutes for the pain to be reduced to less than what would be accomplished using vinegar. But vinegar should not be applied to

Microbiology used in Wine Production

Portuguese man o' war stings, however, since they are not actually jellyfish and vinegar can cause their nematocysts to discharge venom, making the pain worse. Vinegar is often used as a natural deodorant, mainly because of its antibacterial effect.

Contrary to myth, vinegar cannot be used as a detoxification agent to circumvent urinalysis testing for cannabis.

Potential Hazards

Esophageal injury by apple cider vinegar tablets has been reported, and because vinegar products sold for medicinal purposes are neither regulated nor standardized, they vary widely in content, pH, and other respects. Long-term heavy vinegar ingestion may also cause Hypokalemia, Hyperreninemia, and Osteoporosis.

Cervical Screening Tool

Diluted vinegar 3% to 5%, has also been tested as an effective screening tool for cervical cancer. Vinegar changes the colour of affected tissue to white, making diagnosis by inspection possible, reducing in 35% the mortality for early detection against control group.

Cleaning Uses

White vinegar is often used as a household cleaning agent. Because it is acidic, it can dissolve mineral deposits from glass, coffee makers, and other smooth surfaces. For most uses dilution with water is recommended for safety and to avoid damaging the surfaces being cleaned.

Vinegar is an excellent solvent for cleaning epoxy resin and hardener, even after the epoxy has begun to harden. Malt vinegar sprinkled onto crumpled newspaper is a traditional, and still-popular, method of cleaning grease-smeared windows and mirrors in the UK. Vinegar can be used for polishing brass or bronze.

Recently, vinegar has been marketed as a green solution for many household cleaning problems. For example, vinegar has been cited recently as an eco-friendly urine cleaner for pets and as a weed killer.

Agricultural and Horticultural Uses

Herbicide Use

Vinegar can be used as an herbicide. Acetic acid is not absorbed into root systems, the vinegar will kill top growth, but perennial plants will reshoot. Commercial vinegar, available to consumers for household use, does not exceed 5%, and solutions above 10% need careful handling since they are corrosive and damaging to skin. Stronger solutions that are labelled for use as herbicides are available from some retailers.

Miscellaneous

When a bottle of vinegar is opened, mother of vinegar may develop. It is considered harmless and can be removed by filtering.

Vinegar eels (*Turbatrix aceti*), a form of nematode that has cells that are air-borne, may occur in some forms of vinegar unless the vinegar is kept covered. These feed on the mother of vinegar and can occur in naturally fermenting vinegar. This is the reason vinegar condiment jars have tightly-fitting stoppers. Most manufacturers filter and pasteurize their product before bottling to eliminate any potential adulteration, although they are harmless when ingested. When vinegar is added to sodium bicarbonate (baking soda), it produces a volatile mixture which rapidly decomposes into water, carbon dioxide and sodium ethanoate, which makes the reaction fizz. It is often used to illustrate typical acid-base reactions in school science experiments.

Some countries prohibit the selling of vinegar over a certain percentage acidity. As an example, the government of Canada limits the acetic acid of vinegars to between 4.1% and 12.3%.

Posca, a Roman legionaries' basic drink was vinegar mixed with water and optional honey.

According to legend, in France during the Black Plague, four thieves were able to rob houses of plague victims without being infected themselves. When finally caught, the Judge offered to grant the men their freedom, on the condition that they revealed how they managed to stay healthy. They claimed that a medicine woman sold them a potion, made of garlic soaked in soured red wine (vinegar). Variants of the recipe, called Four Thieves Vinegar, have been passed down for hundreds of years and are a staple of New Orleans Voodoo practices.

Diluted vinegar can be used as a homemade stop bath during photographic processing.

Chapter 6
Calcium and Sodium in Food Nutrition

Calcium in Nutrition

Calcium is the chemical element with the symbol Ca and atomic number 20. It has an atomic mass of 40.078 amu. Calcium is a soft gray alkaline earth metal, and is the fifth most abundant element by mass in the Earth's crust. Calcium is also the fifth most abundant dissolved ion in seawater by both molarity and mass, after sodium, chloride, magnesium, and sulfate.

Calcium is essential for living organisms, particularly in cell physiology, where movement of the calcium ion Ca^{2+} into and out of the cytoplasm functions as a signal for many cellular processes. As a major material used in mineralization of bones and shells, calcium is the most abundant metal by mass in many animals.

Notable Characteristics

Chemically calcium is reactive and soft for a metal (though harder than lead, it can be cut with a knife with difficulty). It is a silvery metallic element that must be extracted by electrolysis from a fused salt like calcium chloride. Once produced, it rapidly forms a gray-white oxide and nitride coating when exposed to air. In bulk-form (typically as chips or "turnings") the metal is somewhat difficult to ignite, more so even than magnesium chips; but when lit, the metal burns in air with a brilliant high-intensity red light. Calcium metal reacts with water, evolving hydrogen gas at a rate rapid enough to be noticeable, but not fast enough at room temperature to generate much heat. In powdered form, however, the reaction with water is extremely rapid, as the increased surface area of the powder accelerates the reaction with the water. Part of the slowness of the calcium-water reaction results from the metal being partly protected by insoluble white calcium hydroxide. In water solutions of acids, where this salt is soluble, calcium reacts vigorously.

Calcium, with a density of 1.55 g/cm³, is the lightest of the alkaline earth metals; magnesium (specific gravity 1.74) and beryllium (1.84) are more dense, although lighter in atomic mass. From strontium onward, the alkali earth metals become more dense with increasing atomic mass.

It has two allotropes. Calcium has a higher electrical resistivity than copper or aluminium, yet weight-for-weight, due to its much lower density, it is a rather better conductor than either. However, its use in terrestrial applications is usually limited by its high reactivity with air. Calcium salts are colorless from any contribution of the calcium, and ionic solutions of calcium (Ca^{2+}) are colorless as well. Many calcium salts are not soluble in water. When in solution, the calcium ion to the human taste varies remarkably, being reported as mildly salty, sour, "mineral like" or even "soothing." It is apparent that many animals can taste, or develop a taste, for calcium, and use this sense to detect the mineral in salt licks or other sources. In human nutrition, soluble calcium salts may be added to tart juices without much effect to the average palate.

Calcium is the fifth most abundant element by mass in the human body, where it is a common cellular ionic messenger with many functions, and serves also as a structural element in bone. It is the relatively high atomic-numbered calcium in the skeleton which causes bone to be radio-opaque. Of the human body's solid components after drying and burning of organics (as for example, after cremation), about a third of the total "mineral" mass remaining, is the approximately one kilogram of calcium which composes the average skeleton (the remainder being mostly phosphorus and oxygen).

H and K Lines

Visible spectra of many stars, including the Sun, exhibit strong absorption lines of singly ionized calcium. Prominent among these are the H-line at 3968.5 Å and the K line at 3933.7 Å of singly ionized calcium, or Ca II. For the Sun and stars with low temperatures, the prominence of the H and K lines can be an indication of strong magnetic activity in the chromosphere. Measurement of periodic variations of these active regions can also be used to deduce the rotation periods of these stars.

Compounds

Calcium, combined with phosphate to form hydroxylapatite, is the mineral portion of human and animal bones and teeth. The mineral portion of some corals can also be transformed into hydroxylapatite.

Calcium hydroxide (slaked lime) is used in many chemical refinery processes and is made by heating limestone at high temperature (above 825 °C) and then carefully adding water to it. When lime is mixed with sand, it hardens into a mortar and is turned into plaster by carbon dioxide uptake. Mixed with other compounds, lime forms an important part of Portland cement.

Calcium carbonate ($CaCO_3$) is one of the common compounds of calcium. It is heated to form quicklime (CaO), which is then added to water (H_2O). This forms another material known as slaked lime ($Ca(OH)_2$), which is an inexpensive base material used throughout the chemical industry. Chalk, marble, and limestone are all forms of calcium carbonate.

When water percolates through limestone or other soluble carbonate rocks, it partially dissolves the rock and causes cave formation and characteristic stalactites and stalagmites and also forms hard water. Other important calcium compounds are calcium nitrate, calcium sulfide, calcium chloride, calcium carbide, calcium cyanamide and calcium hypochlorite.

A few calcium compounds in the oxidation state +1 have also been investigated recently.

Nucleosynthesis

Stable Calcium is created in extremely large, extremely hot (over 2.5 billion kelvin) stars. It requires one atom of argon and one atom of helium.

Isotopes

Calcium has four stable isotopes (^{40}Ca and ^{42}Ca through ^{44}Ca), plus two more isotopes (^{46}Ca and ^{48}Ca) that have such long half-lives that for all practical purposes they can be considered stable. The 20% range in relative mass among naturally occurring calcium isotopes is greater than for any element except hydrogen and helium. Calcium also has a cosmogenic isotope, radioactive ^{41}Ca, which has a half-life of 103,000 years. Unlike cosmogenic isotopes that are produced in the atmosphere, ^{41}Ca is produced by neutron activation of ^{40}Ca. Most of its production is in the upper metre or so of the soil column, where the cosmogenic neutron flux is still sufficiently strong. ^{41}Ca has received much attention in stellar studies because it decays to ^{41}K, a critical indicator of solar-system anomalies.

97% of naturally occurring calcium is in the form of ^{40}Ca. ^{40}Ca is one of the daughter products of ^{40}K decay, along with ^{40}Ar. While

K-Ar dating has been used extensively in the geological sciences, the prevalence of ^{40}Ca in nature has impeded its use in dating. Techniques using mass spectrometry and a double spike isotope dilution have been used for K-Ca age dating.

The most abundant isotope, ^{40}Ca, has a nucleus of 20 protons and 20 neutrons. This is the heaviest stable isotope of any element which has equal numbers of protons and neutrons. In supernova explosions, calcium is formed from the reaction of carbon with various numbers of alpha particles (helium nuclei), until the most common calcium isotope (containing 10 helium nuclei) has been synthesized.

Isotope Fractionation

As with the isotopes of other elements, a variety of processes fractionate, or alter the relative abundance of, calcium isotopes. The best studied of these processes is the mass dependent fractionation of calcium isotopes that accompanies the precipitation of calcium minerals, such as calcite, aragonite and apatite, from solution. Isotopically light calcium is preferentially incorporated into minerals, leaving the solution from which the mineral precipitated enriched in isotopically heavy calcium. At room temperature the magnitude of this fractionation is roughly 0.25‰ (0.025%) per atomic mass unit (AMU). Mass-dependant differences in calcium isotope composition conventionally are expressed the ratio of two isotopes (usually ^{44}Ca/^{40}Ca) in a sample compared to the same ratio in a standard reference material. ^{44}Ca/^{40}Ca varies by about 1% among common earth materials.

Calcium isotope fractionation during mineral formation has led to several applications of calcium isotopes. In particular, the 1997 observation by Skulan and DePaolo that calcium minerals are isotopically lighter than the solutions from which the minerals precipitate is the basis of analogous applications in medicine and in paleooceanography. In animals with skeletons mineralized with calcium the calcium isotopic composition of soft tissues reflects the relative rate of formation and dissolution of skeletal mineral. In humans changes in the calcium isotopic composition of urine have been shown to be related to changes in bone mineral balance. When the rate of bone formation exceeds the rate of bone resorption, soft tissue ^{44}Ca/^{40}Ca rises. Soft tissue ^{44}Ca/^{40}Ca falls when bone resorption exceeds bone formation. Because of this relationship, calcium isotopic measurements of urine or blood may be useful in the early detection of metabolic bone diseases like osteoporosis. A similar system exists in the ocean, where seawater ^{44}Ca/^{40}Ca tends to rise when the rate of removal of Ca^{2+} from seawater by mineral precipitation exceeds the input of new calcium into the ocean, and fall when calcium input exceeds mineral precipitation. It follows that rising $^{44}Ca/^{40}Ca$ corresponds to falling seawater Ca^{2+} concentration, and falling $^{44}Ca/^{40}Ca$ corresponds to rising seawater Ca^{2+} concentration. In 1997 Skulan and DePaolo presented the first evidence of change in seawater $^{44}Ca/^{40}Ca$ over geologic time, along with a theoretical explanation of these changes. More recent papers have confirmed this observation, demonstrating that seawater Ca^{2+} concentration is not constant, and that the ocean probably never is in "steady state" with respect to its calcium input and output. This has important climatological implications, as the marine calcium cycle is closely tied to the carbon cycle.

Geochemical Cycling

Calcium provides an important link between tectonics, climate and the carbon cycle. In the simplest terms, uplift of mountains exposes Ca-bearing rocks to chemical weathering and releases Ca^{2+} into surface water. This Ca^{2+} eventually is transported to the ocean where it reacts with dissolved CO_2 to form limestone. Some of this limestone settles to the sea floor where it is incorporated into new rocks. Dissolved CO_2, along with carbonate and bicarbonate ions, are referred to as dissolved inorganic carbon (DIC).

The actual reaction is more complicated and involves the bicarbonate ion (HCO_3^-) that forms when CO_2 reacts with water at seawater pH:

$Ca^{2+} + 2HCO-3 \rightarrow CaCO_3$ (limestone) $+ CO_2 + H_2O$

Note that at ocean pH most of the CO_2 produced in this reaction is immediately converted back into $HCO3^-$. The reaction results in a net transport of one molecule of CO_2 from the ocean/atmosphere into the lithosphere. The result is that each Ca^{2+} ion released by chemical weathering ultimately removes one CO_2 molecule from the surficial system (atmosphere, ocean, soils and living organisms), storing it in carbonate rocks where it is likely to stay for hundreds of millions of years. The weathering of calcium from rocks thus scrubs CO_2 from the ocean and atmosphere, exerting a strong long-term effect on climate. Analogous cycles involving magnesium, and to a much smaller extent strontium and barium, have the same effect.

As the weathering of limestone ($CaCO_3$) liberates equimolar amounts of Ca^{2+} and CO_2, it has no net effect on the CO_2 content of the atmosphere and ocean. The weathering of silicate rocks like granite, on the other hand, is a net CO2 sink because it produces abundant Ca^{2+} very little CO_2.

History

Calcium (from Latin *calx*, genitive *calcis*, meaning "lime") was known as early as the first century when the Ancient Romans prepared lime as calcium oxide. Literature dating back to 975 AD notes that plaster of paris (calcium sulfate), is useful for setting broken bones. It was not isolated until 1808 in England when Sir Humphry Davy electrolyzed a mixture of lime and mercuric oxide. Davy was trying to isolate calcium; when he heard that Swedish chemist Jöns Jakob Berzelius and Pontin prepared calcium amalgam by electrolyzing lime in mercury, he tried it himself. He worked with electrolysis throughout his life and also discovered/isolated sodium, potassium, magnesium, boron and barium. Calcium metal was not available in large scale until the beginning of the 20th century.

Occurrence

See also Category: Calcium minerals Calcium is not naturally found in its elemental state. Calcium occurs most commonly in sedimentary rocks in the minerals calcite, dolomite and gypsum. It also occurs in igneous and metamorphic rocks chiefly in the silicate minerals: plagioclase, amphiboles, pyroxenes and garnets

Applications

Calcium is used

- as a reducing agent in the extraction of other metals, such as uranium, zirconium, and thorium.
- as a deoxidizer, desulfurizer, or decarbonizer for various ferrous and nonferrous alloys.
- as an alloying agent used in the production of aluminium, beryllium, copper, lead, and magnesium alloys.
- in the making of cements and mortars to be used in construction.
- in the making of cheese, where calcium ions influence the activity of rennin in bringing about the coagulation of milk.

Calcium Compounds

- Calcium carbonate ($CaCO_3$) is used in manufacturing cement and mortar, lime, limestone (usually used in the steel industry) and aids in production in the glass industry. It also has chemical and optical uses as mineral specimens in toothpastes, for example.

- Calcium hydroxide solution ($Ca(OH)_2$) (also known as limewater) is used to detect the presence of carbon dioxide by being bubbled through a solution. It turns cloudy where CO_2 is present.
- Calcium arsenate ($Ca_3(AsO_4)_2$) is used in insecticides.
- Calcium carbide (CaC_2) is used to make acetylene gas (for use in acetylene torches for welding) and in the manufacturing of plastics.
- Calcium chloride ($CaCl_2$) is used in ice removal and dust control on dirt roads, in conditioner for concrete, as an additive in canned tomatoes, and to provide body for automobile tires.
- Calcium cyclamate ($Ca(C_6H_{11}NHSO_3)_2$) was used as a sweetening agent but is no longer permitted for use because of suspected cancer-causing properties.
- Calcium gluconate ($Ca(C_6H_{11}O_7)_2$) is used as a food additive and in vitamin pills.
- Calcium hypochlorite ($Ca(OCl)_2$) is used as a swimming pool disinfectant, as a bleaching agent, as an ingredient in deodorant, and in algaecide and fungicide.
- Calcium permanganate ($Ca(MnO_4)_2$) is used in liquid rocket propellant, textile production, as a water sterilizing agent and in dental procedures.
- Calcium phosphate ($Ca_3(PO_4)_2$) is used as a supplement for animal feed, fertilizer, in commercial production for dough and yeast products, in the manufacture of glass, and in dental products.
- Calcium phosphide (Ca_3P_2) is used in fireworks, rodenticide, torpedoes and flares.
- Calcium stearate ($Ca(C_{18}H_{35}O_2)_2$) is used in the manufacture of wax crayons, cements, certain kinds of plastics and cosmetics, as a food additive, in the production of water resistant materials and in the production of paints.
- Calcium sulfate ($CaSO_4 \cdot 2H_2O$) is used as common blackboard chalk, as well as, in its hemihydrate form better known as Plaster of Paris.
- Calcium tungstate ($CaWO_4$) is used in luminous paints, fluorescent lights and in X-ray studies.
- Hydroxylapatite ($Ca_5(PO_4)_3(OH)$, but is usually written $Ca_{10}(PO_4)_6(OH)_2$) makes up seventy percent of bone. Also

carbonated-calcium deficient hydroxylapatite is the main mineral of which dental enamel and dentin are comprised.

Nutrition

Calcium is an important component of a healthy diet and a mineral necessary for life. The National Osteoporosis Foundation says, "Calcium plays an important role in building stronger, denser bones early in life and keeping bones strong and healthy later in life." Approximately ninety-nine percent of the body's calcium is stored in the bones and teeth. The rest of the calcium in the body has other important uses, such as some exocytosis, especially neurotransmitter release, and muscle contraction. In the electrical conduction system of the heart, calcium replaces sodium as the mineral that depolarizes the cell, proliferating the action potential. In cardiac muscle, sodium influx commences an action potential, but during potassium efflux, the cardiac myocyte experiences calcium influx, prolonging the action potential and creating a plateau phase of dynamic equilibrium. Long-term calcium deficiency can lead to rickets and poor blood clotting and in case of a menopausal woman, it can lead to osteoporosis, in which the bone deteriorates and there is an increased risk of fractures. While a lifelong deficit can affect bone and tooth formation, over-retention can cause hypercalcemia (elevated levels of calcium in the blood), impaired kidney function and decreased absorption of other minerals. High calcium intakes or high calcium absorption were previously thought to contribute to the development of kidney stones. However, a high calcium intake has been associated with a lower risk for kidney stones in more recent research. Vitamin D is needed to absorb calcium.

Dairy products, such as milk and cheese, are a well-known source of calcium. Some individuals are allergic to dairy products and even more people, particularly those of non Indo-European descent, are lactose-intolerant, leaving them unable to consume non-fermented dairy products in quantities larger than about half a liter per serving. Others, such as vegans, avoid dairy products for ethical and health reasons. Fortunately, many good sources of calcium exist. These include seaweeds such as kelp, wakame and hijiki; nuts and seeds (like almonds and sesame); blackstrap molasses; beans; oranges; figs; quinoa; amaranth; collard greens; okra; rutabaga; broccoli; dandelion leaves; kale; and fortified products such as orange juice and soy milk. An overlooked source of calcium is eggshell, which can be ground into a powder and mixed into food or a glass of water. Cultivated vegetables generally have less calcium than wild plants.

The calcium content of most foods can be found in the USDA National Nutrient Database.

Calcium in Biology

Calcium (Ca^{2+}) plays a pivotal role in the physiology and biochemistry of organisms and the cell. It plays an important role in signal transduction pathways, where it acts as a second messenger, in neurotransmitter release from neurons, contraction of all muscle cell types, and fertilization. Many enzymes require calcium ions as a cofactor; those of the blood-clotting cascade being notable examples. Extracellular calcium is also important for maintaining the potential difference across excitable cell membranes, as well as proper bone formation.

Calcium levels in mammals are tightly regulated with bone acting as the major mineral storage site. Calcium ions, Ca^{2+}, are released from bone into the bloodstream under controlled conditions. Calcium is transported through the bloodstream as dissolved ions or bound to proteins such as serum albumin.

Parathyroid hormone secreted by the parathyroid gland regulates the resorption of Ca^{2+} from bone, reabsorption in the kidney back into circulation, and increases the activation of vitamin D_3 to Calcitriol. Calcitriol, the active form of vitamin D_3, promotes absorption of calcium from the intestines and the mobilization of calcium ions from bone matrix.

Calcitonin secreted from the parafollicular cells of the thyroid gland also affects calcium levels by opposing parathyroid hormone, however, its physiological significance in humans is dubious.

Calcium Metabolism

Calcium metabolism or calcium homeostasis is the mechanism by which the body maintains adequate calcium levels. Derangements of this mechanism lead to hypercalcemia or hypocalcemia, both of which can have important consequences for health.

Calcium Location and Quantity

Calcium is the most abundant mineral in the human body. The average adult body contains in total approximately 1 kg, 99% in the skeleton in the form of calcium phosphate salts. The extracellular fluid (ECF) contains approximately 22.5 mmol, of which about 9 mmol is in the serum. Approximately 500 mmol of calcium is exchanged between bone and the ECF over a period of twenty-four hours.

Normal Ranges

The serum level of calcium is closely regulated with a normal *total calcium* of 2.2-2.6 mmol/L (9-10.5 mg/dL) and a normal *ionized calcium* of 1.1-1.4 mmol/L (4.5-5.6 mg/dL). The amount of total calcium varies with the level of serum albumin, a protein to which calcium is bound. The biologic effect of calcium is determined by the amount of *ionized calcium*, rather than the total calcium. Ionized calcium does not vary with the albumin level, and therefore it is useful to measure the ionized calcium level when the serum albumin is not within normal ranges, or when a calcium disorder is suspected despite a normal total calcium level.

Corrected Calcium Level

One can derive a corrected calcium level when the albumin is abnormal. This is to make up for the change in total calcium due to the change in albumin-bound calcium, and gives an estimate of what the calcium level would be if the albumin were within normal ranges.

Corrected calcium (mg/dL) = measured total Ca (mg/dL) + 0.8 (4.0 -serum albumin [g/dL]), where 4.0 represents the average albumin level in g/dL. in other words, each 1 g/dL decrease of albumin will decrease 0.8 mg/dL in measured serum Ca and thus 0.8 must be added to the measured Calcium to get a corrected Calcium value.

Or: Corrected calcium (mmol/L) = measured total Ca (mmol/L) + 0.02 (40 -serum albumin [g/L]), where 40 represents the average albumin level in g/L in other words, each 1 g/L decrease of albumin, will decrease 0.02 mmol/L in measured serum Ca and thus 0.02 must be added to the measured value to take this into account and get a corrected calcium.

When there is hypoalbuminemia (a lower than normal albumin), the corrected calcium level is higher than the total calcium.

Effector Organs

Sources

About 25 mmol of calcium enters the body in a normal diet. Of this, about 40% (10 mmol) is absorbed in gut, and 5 mmol leaves the body in feces, netting 5 mmol of calcium a day.

Excretion

The kidney excretes 250 mmol a day in pro-urine, and resorbs 245 mmol, leading to a net loss in the urine of 5 mmol/d. In addition to this, the kidney processes Vitamin D into calcitriol, the active form that is most effective in assisting intestinal absorption. Both processes are stimulated by parathyroid hormone.

The Role of Bone

Although calcium flow to and from the bone is neutral, about 5 mmol is turned over a day. Bone serves as an important storage point for calcium, as it contains 99% of the total body calcium. Calcium release from bone is regulated by parathyroid hormone. Calcitonin stimulates incorporation of calcium in bone, although this process is largely independent of calcitonin.

Low calcium intake may also be a risk factor in the development of osteoporosis. In one meta-analysis, the authors found that fifty out of the fifty-two studies that they reviewed showed that calcium intake promoted better bone balance. With a better bone balance, the risk of osteoporosis is lowered.

Interaction with Other Chemicals

Potential Positive Interactions

- Vitamin D is an important co-factor in the intestinal absorption of calcium, as it increases the number of calcium binding proteins, involved in calcium absorption through the apical membrane of enterocytes in small intestine. It also promotes re-absorption of calcium in the kidneys.
- Boron.

Potential Negative Interactions

- "Unesterified long-chain saturated fatty acids, i.e. palmitic acid, have a melting point above body temperature and, with sufficient calcium in the intestinal lumen, form insoluble calcium soaps."
- Sodium binding to calcium
- Phytic acid binding to calcium
- Oxalic acid binding to calcium
- Caffeine binding to calcium
- Cortisol binding to calcium
- Low pH food and proteins (the latter promotes gastric acid).

Regulatory Organs

Primarily calcium is regulated by the actions of 1,25-OH-vitamin D_3, parathyroid hormone (PTH) and calcitonin and direct exchange with the bone matrix. Plasma calcium levels are regulated by hormonal and non-hormonal mechanisms. After ingestion of substantial amounts of calcium, for example in a glass of milk, the short term control that prevents calcium spiking in the serum is absorption by the bone matrix. After about an hour, PTH will be released and not peak for about 8 hours. The PTH is, over time, a very potent regulator of plasma calcium, and controls the conversion of vitamin D into its active form in the kidney. The parathyroid glands are located behind the thyroid, and produce parathyroid hormone in response to low calcium levels.

The parafollicular cells of the thyroid produce calcitonin in response to high calcium levels, but its significance is much smaller than that of PTH.

Pathology

Hypocalcemia and hypercalcemia are both serious medical disorders.

Renal osteodystrophy is a consequence of chronic renal failure related to the calcium metabolism.

Osteoporosis and osteomalacia have been linked to calcium metabolism disorders.

Research into Cancer Prevention

The role that calcium might have in reducing the rates of colorectal cancer has been the subject of many studies. However, given its modest efficacy, there is no current medical recommendation to use calcium for cancer reduction. Several epidemiological studies suggest that people with high calcium intake have a reduced risk of colorectal cancer. These observations have been confirmed by experimental studies in volunteers and in rodents. One large scale clinical trial shows that 1.2 g calcium each day reduces, modestly, intestinal polyps recurrence in volunteers. Data from the four published trials are available. Some forty carcinogenesis studies in rats or mice, reported in the Chemoprev. Database, also support that calcium could prevent intestinal cancer.

Dietary Calcium Supplements

Calcium supplements are used to prevent and to treat calcium deficiencies. Most experts recommend that supplements be taken with food and that no more than 600 mg should be taken at a time because the percent of calcium absorbed decreases as the amount of calcium in the supplement increases. It is recommended to spread doses throughout the day. Recommended daily calcium intake for adults ranges from 1000 to 1500 mg. It is recommended to take supplements with food to aid in absorption.

Vitamin D is added to some calcium supplements. Proper vitamin D status is important because vitamin D is converted to a hormone in the body which then induces the synthesis of intestinal proteins responsible for calcium absorption.

- The absorption of calcium from most food and commonly used dietary supplements is very similar. This is contrary to what many calcium supplement manufacturers claim in their promotional materials.
- Milk is an excellent source of dietary calcium for those whose body tolerate it because it has a high concentration of calcium and the calcium in milk is excellently absorbed.
- Soymilk and other vegetable milks are usually sold with calcium added so that their calcium concentration is as high as in milk
- Also different kind of juices boosted with calcium are widely available.
- Calcium carbonate is the most common and least expensive calcium supplement. It should be taken with food. It depends on low pH levels for proper absorption in the intestine. Some studies suggests that the absorption of calcium from calcium carbonate is similar to the absorption of calcium from milk. While most people digest calcium carbonate very well, some might develop gastrointestinal discomfort or gas. Taking magnesium with it can help to avoid constipation. Calcium carbonate is 40% elemental calcium. 1000 mg will provide 400 mg of calcium. However, supplement labels will usually indicate how much calcium is present in each serving, not how much calcium carbonate is present.
- Antacids frequently contain calcium carbonate, and are a commonly used, inexpensive calcium supplement
- Coral Calcium is a salt of calcium derived from fossilized coral reefs. Coral calcium is composed of calcium carbonate and trace minerals.
- Calcium citrate can be taken without food and is the supplement of choice for individuals with achlorhydria or who are taking histamine-2 blockers or proton-pump inhibitors. It is more

easily digested and absorbed than calcium carbonate if taken on an empty stomach and less likely to cause constipation and gas than calcium carbonate. It also has a lower risk of contributing to the formation of kidney stones. Calcium citrate is about 21% elemental calcium. 1000 mg will provide 210 mg of calcium. It is more expensive than calcium carbonate and more of it must be taken to get the same amount of calcium.

- Calcium phosphate costs more than calcium carbonate, but less than calcium citrate. It is easily absorbed and is less likely to cause constipation and gas than either.
- Calcium lactate has similar absorption as calcium carbonate, but is more expensive. Calcium lactate and calcium gluconate are less concentrated forms of calcium and are not practical oral supplements.
- Calcium chelates are synthetic calcium compounds, with calcium bound to an organic molecule, such as malate, aspartate, or fumarate. These forms of calcium may be better absorbed on an empty stomach. However, in general they are absorbed similarly to calcium carbonate and other common calcium supplements when taken with food. The 'chelate' mimics the action that natural food performs by keeping the calcium soluble in the intestine. Thus, on an empty stomach, in some individuals, chelates might theoretically be absorbed better.
- Microcrystalline hydroxyapatite (MH) is marketed as a calcium supplement, and has in some randomized trials been found to be more effective than calcium carbonate.

In July 2006, a report citing research from Fred Hutchinson Cancer Research Centre in Seattle, Washington claimed that women in their 50s gained 5 pounds less in a period of 10 years by taking more than 500 mg of calcium supplements than those who did not. However, the doctor in charge of the study, Dr. Alejandro J. Gonzalez also noted it would be "going out on a limb" to suggest calcium supplements as a weight-limiting aid.

Prevention of Fractures due to Osteoporosis

Such studies often do not test calcium alone, but rather combinations of calcium and vitamin D. Randomized controlled trials found both positive and negative effects. The different results may be explained by doses of calcium and underlying rates of calcium supplementation in the control groups. However, it is clear that increasing the intake of calcium promotes deposition of calcium in the bones, where it is of more benefit in preventing the compression fractures resulting from the osteoporotic thinning of the dendritic web of the bodies of the vertebrae, than it is at preventing the more serious cortical bone fractures which happen at hip and wrist.

Possible Cancer Prevention

A meta-analysis by the international Cochrane Collaboration of two randomized controlled trials found that calcium "might contribute to a moderate degree to the prevention of adenomatous colonic polyps".

More recent studies were conflicting, and one which was positive for effect (Lappe, et al.) did control for a possible anti-carcinogenic effect of vitamin D, which was found to be an independent positive influence from calcium-alone on cancer risk.

- A randomized controlled trial found that 1000 mg of elemental calcium and 400 IU of vitamin D_3 had no effect on colorectal cancer
- A randomized controlled trial found that 1400–1500 mg supplemental calcium and 1100 IU vitamin D_3 reduced aggregated cancers with a relative risk of 0.402.
- An observational cohort study found that high calcium and vitamin D intake was associated with "lower risk of developing premenopausal breast cancer."

Hazards and Toxicity

Compared to other metals, the calcium ion and most calcium compounds have low toxicity. This is not surprising given the very high natural abundance of calcium compounds in the environment and in organisms. Calcium poses few, if any, serious environmental problems. Acute calcium poisoning is rare, and difficult to achieve unless calcium compounds are administered intravenously. For example, the oral median lethal dose (LD^{50}) for rats for calcium carbonate and calcium chloride are 6.45 and 1.4 g/kg, respectively.

Calcium metal is hazardous because of its sometimes violent reactions with water and acids. Calcium metal is found in some drain cleaners, where it functions to generate heat and calcium hydroxide that saponifies the fats and liquefies the proteins (e.g., hair) that block drains. When swallowed calcium metal has the same effect on the mouth, esophagus and stomach, and can be fatal.

Excessive consumption of calcium carbonate antacids/dietary supplements (such as Tums) over a period of weeks or months can cause milk-alkali syndrome, with symptoms ranging from hypercalcemia to potentially fatal renal failure. What constitutes "excessive" consumption is not well known and probably varies a great deal from person to person. Persons who consume more than 10 grams/day of $CaCO_3$ (=4 g Ca) are at risk of developing milk-alkali syndrome, but the condition has been reported in at least one person consuming only 2.5 grams/day of $CaCO_3$ (=1 g Ca), an amount usually considered moderate and safe.

Sodium in Nutrition

Sodium is a metallic element with a symbol Na and atomic number 11. It is a soft, silvery-white, highly reactive metal and is a member of the alkali metals within "group 1" (formerly known as 'group IA'). It has only one stable isotope, ^{23}Na.

Elemental sodium was first isolated by Humphry Davy in 1807 by passing an electric current through molten sodium hydroxide. Elemental sodium does not occur naturally on Earth, because it quickly oxidizes in air and is violently reactive with water, so it must be stored in an inert medium, such as a liquid hydrocarbon. The free metal is used for some chemical synthesis, analysis, and heat transfer applications. Sodium ion is soluble in water in nearly all of its compounds, and is thus present in great quantities in the Earth's oceans and other stagnant bodies of water. In these bodies it is mostly counterbalanced by the chloride ion, causing evaporated ocean water solids to consist mostly of sodium chloride, or common table salt. Sodium ion is also a component of many minerals.

Sodium is an essential element for all animal life (including human) and for some plant species. In animals, sodium ions are used in opposition to potassium ions, to allow the organism to build up an electrostatic charge on cell membranes, and thus allow transmission of nerve impulses when the charge is allowed to dissipate by a moving wave of voltage change. Sodium is thus classified as a "dietary inorganic macro-mineral" for animals. Sodium's relative rarity on land is due to its solubility in water, thus causing it to be leached into bodies of long-standing water by rainfall. Such is its relatively large requirement in animals, in contrast to its relative scarcity in many inland soils, that herbivorous land animals have developed a special taste receptor for the sodium ion.

Characteristics

At room temperature, sodium metal is soft enough that it can be cut with a knife. In air, the bright silvery luster of freshly exposed sodium will rapidly tarnish. The density of alkali metals generally increases with increasing atomic number, but sodium is denser than potassium. Sodium is a fairly good conductor of heat.

Chemical Properties

Compared with other alkali metals, sodium is generally less reactive than potassium and more reactive than lithium, in accordance with "periodic law": for example, their reaction in water, chlorine gas, etc.

Sodium reacts exothermically with water: small pea-sized pieces will bounce across the surface of the water until they are consumed by it, whereas large pieces will explode. While sodium reacts with water at room temperature, the sodium piece melts with the heat of the reaction to form a sphere, if the reacting sodium piece is large enough. The reaction with water produces very caustic sodium hydroxide (lye) and highly flammable hydrogen gas. These are extreme hazards. When burned in air, sodium forms sodium peroxide Na_2O_2, or with limited oxygen, the oxide Na_2O (unlike lithium, the nitride is not formed). If burned in oxygen under pressure, sodium superoxide NaO_2 will be produced.

Compounds

Sodium compounds are important to the chemical, glass, metal, paper, petroleum, pyrotechnic, soap, and textile industries. Hard soaps are generally sodium salt of certain fatty acids (potassium produces softer or liquid soaps).

The sodium compounds that are the most important to industries are common salt (NaCl), soda ash (Na_2CO_3), baking soda ($NaHCO_3$), caustic soda (NaOH), sodium nitrate ($NaNO_3$), di-and tri-sodium phosphates, sodium thiosulfate ($Na_2S_2O_3 \cdot 5H_2O$), and borax ($Na_2B_4O_7 \cdot 10H_2O$).

Sodium tends to form water-soluble compounds, such as halides, sulfate, nitrate, carboxylates and carbonates. There are only isolated examples of sodium compounds precipitating from water solution. However, nature provides examples of many insoluble sodium compounds such as the feldspars (aluminum silicates of sodium, potassium and calcium). There are other insoluble sodium salts such as sodium bismuthate $NaBiO_3$, sodium octamolybdate $Na_2Mo_8O_{25} \cdot 4H_2O$, sodium thioplatinate $Na_4Pt_3S_6$, sodium uranate Na_2UO_4. Sodium meta-antimonate's $2NaSbO_3 \cdot 7H_2O$ solubility is 0.3 g/L as is the pyro form $Na_2H_2Sb_2O_7 \cdot H_2O$ of this salt. Sodium metaphosphate $NaPO_3$ has a soluble and an insoluble form.

Spectroscopy

When sodium or its compounds are introduced into a flame, they turn the flame a bright yellow colour.

One notable atomic spectral line of sodium vapour is the so-called D-line, which may be observed directly as the sodium flame-test line and also the major light output of low-pressure sodium lamps (these produce an unnatural yellow, rather than the peach-coloured glow of high pressure lamps). The D-line is one of the classified Fraunhofer lines observed in the visible spectrum of the Sun's electromagnetic radiation. Sodium vapour in the upper layers of the Sun creates a dark line in the emitted spectrum of electromagnetic radiation by absorbing visible light in a band of wavelengths around 589.5 nm. This wavelength corresponds to transitions in atomic sodium in which the valence-electron transitions from a 3p to 3s electronic state. Closer examination of the visible spectrum of atomic sodium reveals that the D-line actually consists of two lines called the D_1 and D_2 lines at 589.6 nm and 589.0 nm, respectively. This fine structure results from a spin-orbit interaction of the valence electron in the 3p electronic state. The spin-orbit interaction couples the spin angular momentum and orbital angular momentum of a 3p electron to form two states that are respectively notated as 3p(^{2}P01/2) and 3p(^{2}P03/2) in the LS coupling scheme. The 3s state of the electron gives rise to a single state which is notated as 3s($^2S_{1/2}$) in the LS coupling scheme. The D_1-line results from an electronic transition between 3s($^2S_{1/2}$) lower state and 3p(^{2}P01/2) upper state. The D_2-line results from an electronic transition between 3s($^2S_{1/2}$) lower state and 3p(^{2}P03/2) upper state. Even closer examination of the visible spectrum of atomic sodium would reveal that the D-line actually consists of a lot more than two lines. These lines are associated with hyperfine structure of the 3p upper states and 3s lower states. Many different transitions involving visible light near 589.5 nm may occur between the different upper and lower hyperfine levels.

A practical use for lasers which work at the sodium D-line transition is to create artificial laser guide stars (artificial star-like images from sodium in the upper atmosphere) which assist in the adaptive optics for large land-based visible light telescopes.

Isotopes

Nearly twenty isotopes of sodium have been recognized, the only stable one being ^{23}Na. Sodium has two radioactive cosmogenic isotopes which are also the two isotopes with longest half life, ^{22}Na, with a half-life of 2.6 years and ^{24}Na with a half-life of 15 hours. All other isotopes have a half life of less than one minute. Acute neutron radiation exposure (e.g., from a nuclear criticality accident) converts some of the stable ^{23}Na in human blood plasma to ^{24}Na. By measuring the concentration of this isotope, the neutron radiation dosage to the victim can be computed.

History

Salt has been an important commodity in human activities, as testified by the English word *salary*, referring to *salarium*, the wafers of salt sometimes given to Roman soldiers along with their other wages. In medieval Europe a compound of sodium with the Latin name of *sodanum* was used as a headache remedy. The name sodium probably originates from the Arabic word suda meaning headache as the headache-alleviating properties of sodium carbonate or soda were well known in early times.

Sodium's chemical abbreviation *Na* was first published by Jöns Jakob Berzelius in his system of atomic symbols (Thomas Thomson, *Annals of Philosophy*) and is a contraction of the element's new Latin name *natrium* which refers to the Egyptian *natron*, the word for a natural mineral salt whose primary ingredient is hydrated sodium carbonate. Hydrated sodium carbonate historically had several important industrial and household uses later eclipsed by soda ash, baking soda and other sodium compounds.

Although sodium (sometimes called "soda" in English) has long been recognized in compounds, it was not isolated until 1807 by Humphry Davy through the electrolysis of caustic soda.

Sodium imparts an intense yellow colour to flames. As early as 1860, Kirchhoff and Bunsen noted the high sensitivity that a flame test for sodium could give. They state in Annalen der Physik und Chemie in the paper "Chemical Analysis by Observation of Spectra":

In a corner of our 60 m³ room farthest away from the apparatus, we exploded 3 mg. of sodium chlorate with milk sugar while observing the nonluminous flame before the slit. After a while, it glowed a bright yellow and showed a strong sodium line that disappeared only after 10 minutes. From the weight of the sodium salt and the volume of air in the room, we easily calculate that one part by weight of air could not contain more than 1/20 millionth weight of sodium.

Creation

Stable forms of sodium are created in stars through nuclear fusion by fusing two carbon atoms together. This requires temperatures above 600 megakelvins, and a large star with at least three solar masses.

Occurrence

Owing to its high reactivity, sodium is found in nature only as a compound and never as the free element. Sodium makes up about 2.6% by weight of the Earth's crust, making it the sixth most abundant element overall and the most abundant alkali metal. Sodium is found in many different minerals, of which the most common is ordinary salt (sodium chloride), which occurs in vast quantities dissolved in seawater, as well as in solid deposits (halite). Others include amphibole, cryolite, soda niter and zeolite.

Sodium is relatively abundant in stars and the D spectral lines of this element are among the most prominent in star light. Though elemental sodium has a rather high vaporization temperature, its relatively high abundance and very intense spectral lines have allowed its presence to be detected by ground telescopes and confirmed by spacecraft (Mariner 10 and MESSENGER) in the thin atmosphere of the planet Mercury.

Commercial Production

Sodium was first produced commercially in 1855 by thermal reduction of sodium carbonate with carbon at 1100 °C, in what is known as the Deville process.

$$Na_2CO_3 \text{ (l)} + 2 \text{ C (s)} \rightarrow 2 \text{ Na (g)} + 3 \text{ CO (g)}$$

A process based on the reduction of sodium hydroxide was developed in 1886.

Sodium is now produced commercially through the electrolysis of liquid sodium chloride, based on a process patented in 1924. This is done in a Downs Cell in which the NaCl is mixed with calcium chloride to lower the melting point below 700 °C. As calcium is less electropositive than sodium, no calcium will be formed at the anode. This method is less expensive than the previous Castner process of electrolyzing sodium hydroxide.

Very pure sodium can be isolated by the thermal decomposition of sodium azide.

Sodium metal in reagent-grade sold for about $1.50/pound ($3.30/kg) in 2009 when purchased in tonnage quantities. Lower purity metal sells for considerably less. The market in this metal is volatile due to the difficulty in its storage and shipping. It must be stored under a dry inert gas atmosphere or anhydrous mineral oil to prevent the formation of a surface layer of sodium oxide or sodium superoxide. These oxides can react violently in the presence of organic materials. Sodium will also burn violently when heated in air.

Smaller quantities of sodium, such as a kilogram, cost far more, in the range of $165/kg. This is partially due to the cost of shipping hazardous material.

Applications

Metallic Sodium

- Sodium in its metallic form can be used to refine some reactive metals, such as zirconium and potassium, from their compounds.
- In certain alloys to improve their structure.
- To descale metal (make its surface smooth).
- To purify molten metals.
- In sodium vapour lamps, an efficient means of producing light from electricity, often used for street lighting in cities. Low-pressure sodium lamps give a distinctive yellow-orange light which consists primarily of the twin sodium D lines. High-pressure sodium lamps give a more natural peach-coloured light, composed of wavelengths spread much more widely across the spectrum.
- As a heat transfer fluid in some types of nuclear reactors and inside the hollow valves of high-performance internal combustion engines.
- In organic synthesis, sodium is used as a reducing agent, for example in the Birch reduction.
- In chemistry, sodium is often used either alone or with potassium in an alloy, NaK as a desiccant for drying solvents. Used with benzophenone, it forms an intense blue coloration when the solvent is dry and oxygen-free.
- The sodium fusion test uses sodium's high reactivity, low melting point, and the near-universal solubility of its compounds, to qualitatively analyze compounds.

Nuclear Reactor Cooling

Molten sodium is used as a coolant in some types of fast neutron reactors. It has a low neutron absorption cross section, which is required to achieve a high enough neutron flux, and has excellent thermal conductivity. Its high boiling point allows the reactor to operate at ambient pressure. However, using sodium poses certain challenges. The molten metal will readily burn in air and react violently with water, liberating explosive hydrogen. During reactor operation, a small amount of sodium-24 is formed as a result of neutron activation, making the coolant radioactive.

Sodium leaks and fires were a significant operational problem in the first large sodium-cooled fast reactors, causing extended shutdowns at the Monju Nuclear Power Plant and Beloyarsk Nuclear Power Plant.

Where reactors need to be frequently shut down, as is the case with some research reactors, the alloy of sodium and potassium called NaK is used. It melts at "11°C, so cooling pipes will not freeze at room temperature. Extra precautions against coolant leaks need to be taken in case of NaK, because molten potassium will spontaneously catch fire when exposed to air.

Sodium Compounds

- This alkali metal as the Na^+ ion is vital to animal life.
- In soap, as sodium salts of fatty acids. Sodium soaps are harder (higher melting) soaps than potassium soaps.
- In some medicine formulations, the salt form of the active ingredient usually with sodium or potassium is a common modification to improve bioavailability.
- Sodium chloride (NaCl), a compound of sodium ions and chloride ions, is an important heat transfer material.

Biological Role

Maintaining Body Fluid Volume in Animals

The serum sodium and urine sodium play important roles in medicine, both in the maintenance of sodium and total body fluid homeostasis, and in the diagnosis of disorders causing homeostatic disruption of salt/sodium and water balance.

In mammals, decreases in blood pressure and decreases in sodium concentration sensed within the kidney result in the production of renin, a hormone which acts in a number of ways, one of them being to act indirectly to cause the generation of aldosterone, a hormone which decreases the excretion of sodium in the urine. As the body of the mammal retains more sodium, other osmoregulation systems which sense osmotic pressure in part from the concentration of sodium and water in the blood, act to generate antidiuretic hormone. This, in turn, causes the body to retain water, thus helping to restore the body's total amount of fluid.

There is also a counterbalancing system, which senses volume. As fluid is retained, receptors in the heart and vessels which sense distension and pressure, cause production of atrial natriuretic peptide, which is named in part for the Latin word for sodium. This hormone acts in various ways to cause the body to lose sodium in the urine. This causes the body's osmotic balance to drop (as low concentration of sodium is sensed directly), which in turn causes the osmoregulation system to excrete the "excess" water. The net effect is to return the body's total fluid levels back toward normal.

Maintaining Electric Potential in Animal Tissues

Sodium cations are important in neuron (brain and nerve) function, and in influencing osmotic balance between cells and the interstitial fluid, with their distribution mediated in all animals (but not in all plants) by the so-called Na^+/K^+-ATPase pump. Sodium is the chief cation in fluid residing outside cells in the mammalian body (the so-called extracellular compartment), with relatively little sodium residing inside cells.

The volume of extracellular fluid is typically 15 litres in a 70 kg human, and the 50 grams of sodium it contains is about 90% of the body's total sodium content.

Role of Sodium in Plant Biology

Although sodium is not considered an essential micronutrient in most plants, it is necessary in the metabolism of some C4 plants, e.g. Rhodes grass, amaranth, Joseph's coat, and pearl millet.

Within these C4 plants, sodium is used in the regeneration of phosphoenolpyruvate (PEP) and the synthesis of chlorophyll. In addition, the presence of sodium can offset potassium requirements in many plants by substituting in several roles, such as: maintaining turgor pressure, serving as an accompanying cation in long distance transport, and aiding in stomatal opening and closing.

Due to increasing soil salinity, osmotic stress and sodium toxicity in plants, especially in agricultural crops, have become worldwide phenomena. High levels of sodium in the soil solution limit the plants' ability to uptake water due to decreased soil water potential and, therefore, may result in wilting of the plant. In addition, excess sodium within the cytoplasm of plant cells can lead to enzyme inhibition, which may result in symptoms such as necrosis, chlorosis, and possible plant death.

To avoid such symptoms, plants have developed methods to combat high sodium levels, such as: mechanisms limiting sodium uptake by roots, compartmentalization of sodium in cell vacuoles, and control of sodium in long distance transport. Many plants store excess sodium in old plant tissue, limiting damage to new growth.

Dietary Uses

The most common sodium salt, sodium chloride ('table salt' or 'common salt'), is used for seasoning and warm-climate food preservation, such as pickling and making jerky (the high osmotic content of salt inhibits bacterial and fungal growth).

The human requirement for sodium in the diet is about 1.5 grams per day, which is typically less than a tenth as much as many diets "seasoned to taste." Most people consume far more sodium than is physiologically needed. Low sodium intake may lead to sodium deficiency (hyponatremia).

Persons suffering from severe dehydration caused by diarrhea, such as that by cholera, can be treated with oral rehydration therapy, in which they drink a solution of sodium chloride, potassium chloride and glucose. This simple, effective therapy saves the lives of millions of children annually in the developing world.

Salt

Salt, also known as table salt, or rock salt, is a mineral that is composed primarily of sodium chloride. It is essential for animal life in small quantities, but is harmful to animals and plants in excess. Salt is one of the oldest, most ubiquitous food seasonings and salting is an important method of food preservation. The taste of salt (saltiness) is one of the basic human tastes.

Salt for human consumption is produced in different forms: unrefined salt (such as sea salt), refined salt (table salt), and iodized salt. It is a crystalline solid, white, pale pink or light gray in colour, normally obtained from sea water or rock deposits. Edible rock salts may be slightly grayish in colour because of mineral content.

Chloride and sodium ions, the two major components of salt, are needed by all known living creatures in small quantities. Salt is involved in regulating the water content (fluid balance) of the body.

However, too much salt increases the risk of health problems, including high blood pressure. Therefore health authorities have recommended limitations of dietary sodium.

While people have used canning and artificial refrigeration to preserve food for the last hundred years or so, salt has been the best-known food preservative, especially for meat, for many thousands of years. A very ancient saltworks operation has been discovered at the Poiana Slatinei archaeological site next to a salt spring in Lunca, Neamţ County, Romania.

Evidence indicates that Neolithic people of the Precucuteni Culture were boiling the salt-laden spring water through the process of briquetage to extract the salt as far back as 6050 BC. The salt extracted from this operation may have had a direct correlation to the rapid growth of this society's population soon after its initial production began. The harvest of salt from the surface of Xiechi Lake near Yuncheng in Shanxi, China dates back to at least 6000 BC, making it one of the oldest verifiable saltworks.

Salt was included among funereal offerings found in ancient Egyptian tombs from the third millennium BC, as were salted birds and salt fish. From about 2800 BC, the Egyptians began exporting salt fish to the Phoenicians in return for Lebanon cedar, glass, and the dye Tyrian purple; the Phoenicians traded Egyptian salt fish and salt from North Africa throughout their Mediterranean trade empire.

Along the Sahara, the Tuareg maintain routes especially for the transport of salt by Azalai (salt caravans). In 1960, the caravans still transported some 15,000 tons of salt, but this trade has now declined to roughly a third.

Salzburg, Hallstatt, and Hallein lie on the river Salzach in central Austria, within a radius of no more than 17 kilometres. Salzach literally means "salt water" and Salzburg "salt city", both taking their names from the German word for salt, Salz.

Hallstatt gave its name to the Celtic archaeological culture that began mining for salt in the area in around 800 BC. Around 400 BC, the Hallstatt Celts, who had heretofore mined for salt, began open pan salt making. During the first millennium BC, Celtic communities grew rich trading salt and salted meat to Ancient Greece and Ancient Rome in exchange for wine and other luxuries.

It is widely, though incorrectly, believed that troops in the Roman army were paid in salt. Even widely respected historical works repeat this error. The word *salad* literally means "salted," and comes from the ancient Roman practice of salting leaf vegetables.

Mahatma Gandhi led at least 100,000 people on the "Dandi March" or "Salt Satyagraha", in which protesters made their own salt from the sea, which was illegal under British rule, as it avoided paying the "salt tax". This civil disobedience inspired millions of common people, and elevated the Indian independence movement from an elitist struggle to a national struggle.

Forms of Salt

Unrefined Salt

Different natural salts have different mineralities, giving each one a unique flavour. Fleur de sel, natural sea salt harvested by hand, has a unique flavour varying from region to region. In traditional Korean cuisine, so-called "bamboo salt" is prepared by roasting salt in a bamboo container plugged with mud at both ends. This product absorbs minerals from the bamboo and the mud, and has been shown to increase the anticlastogenic and antimutagenic properties of doenjang.

Completely raw sea salt is bitter because of magnesium and calcium compounds, and thus is rarely eaten. The refined salt industry cites scientific studies saying that raw sea and rock salts do not contain enough iodine salts to prevent iodine deficiency diseases.

Unrefined sea salts are also commonly used as ingredients in bathing additives and cosmetic products. One example is bath salts, which uses sea salt as its main ingredient and combined with other ingredients used for its healing and therapeutic effects.

Refined Salt

Refined salt, which is most widely used presently, is mainly sodium chloride. Food grade salt accounts for only a small part of salt production in industrialised countries (3% in Europe) although worldwide, food uses account for 17.5% of salt production. The majority is sold for industrial use. Salt has great commercial value because it is a necessary ingredient in the manufacturing of many things. A few common examples include: the production of pulp and paper, setting dyes in textiles and fabrics, and the making of soaps and detergents.

The manufacture and use of salt is one of the oldest chemical industries. Salt can be obtained by evaporation of sea water, usually in shallow basins warmed by sunlight; salt so obtained was formerly called bay salt, and is now often called sea salt or solar salt. Rock salt deposits are formed by the evaporation of ancient salt lakes, and may be mined conventionally or through the injection of water. Injected water dissolves the salt, and the brine solution can be pumped to the surface where the salt is collected.

After the raw salt is obtained, it is refined to purify it and improve its storage and handling characteristics. Purification usually involves recrystallization. In recrystallization, a brine solution is treated with chemicals that precipitate most impurities (largely magnesium and calcium salts).

Multiple stages of evaporation are then used to collect pure sodium chloride crystals, which are kiln-dried. Since the 1950s it has been common practice in the United Kingdom to add a trace amount of sodium ferrocyanide to the brine; this acts as an anticaking agent by promoting irregular crystals.

The safety of sodium ferrocyanide as a food additive was confirmed in the United Kingdom in 1993. Some anticaking agents used are tricalcium phosphate, calcium or magnesium carbonates, fatty acid salts (acid salts), magnesium oxide, silicon dioxide, calcium silicate, sodium aluminosilicate, and calcium aluminosilicate. Both the European Union and the United States Food and Drug Administration (FDA) permitted the use of aluminum in the latter two compounds. The refined salt is then ready for packing and distribution.

Table Salt

In Western cuisines, salt is used in cooking, and also made available to diners in salt shakers on the table.

Table salt is refined salt, which contains about 97% to 99% sodium chloride. It usually contains substances that make it free-flowing (anticaking agents) such as sodium silicoaluminate or magnesium carbonate. Some people also add a desiccant, such as a few grains of uncooked rice, in salt shakers to absorb extra moisture and help break up clumps when anticaking agents are not enough. Table salt has a particle density of 2.165 g/cm^3, and a bulk density (dry, ASTM D 632 gradation) of about 1.154 g/cm^3.

Salty Condiments

In many East Asian cultures, salt is not traditionally used as a condiment. However, condiments such as soy sauce, fish sauce and oyster sauce tend to have a high salt content and fill much the same role as a salt-providing table condiment that table salt serves in western cultures.

Health Effects

Sodium is one of the primary electrolytes in the body. All four cationic electrolytes (sodium, potassium, magnesium, and calcium) are available in unrefined salt, as are other vital minerals needed for optimal bodily function. Too much or too little salt in the diet can lead to muscle cramps, dizziness, or electrolyte disturbance, which can cause neurological problems, or death. Drinking too much water, with insufficient salt intake, puts a person at risk of water intoxication (hyponatremia). Salt is sometimes used as a health aid, such as in treatment of dysautonomia.

Excess salt consumption is linked with a number of conditions including :

- Stroke and cardiovascular disease.
- Hypertension (high blood pressure): "Since 1994, the evidence of an association between dietary salt intakes and blood pressure has increased. The data have been consistent in various study populations and across the age range in adults." A large scale study from 2007 has shown that people with high-normal blood pressure who significantly reduced the amount of salt in their diet decreased their chances of developing cardiovascular disease by 25% over the following 10 to 15 years. Their risk of dying from cardiovascular disease decreased by 20%.
- Left ventricular hypertrophy (cardiac enlargement): "Evidence suggests that high salt intake causes left ventricular hypertrophy, a strong risk factor for cardiovascular disease, independently of blood pressure effects." "...there is accumulating evidence that high salt intake predicts left ventricular hypertrophy." Excessive salt (sodium) intake, combined with an inadequate intake of water, can cause hypernatremia. It can exacerbate renal disease.
- Edema (BE: oedema): A decrease in salt intake has been suggested to treat edema (fluid retention).
- Duodenal ulcers and gastric ulcers
- Heartburn.
- Osteoporosis: One report shows that a high salt diet does reduce bone density in women. Yet "While high salt intakes

have been associated with detrimental effects on bone health, there are insufficient data to draw firm conclusions."
- Gastric cancer (stomach cancer) is associated with high levels of sodium, "but the evidence does not generally relate to foods typically consumed in the UK." However, in Japan, salt consumption is higher.
- Death: Ingestion of large amounts of salt in a short time (about 1 g per kg of body weight) can be fatal. Deaths have also resulted from attempted use of salt solutions as emetics, forced salt intake, and accidental confusion of salt with sugar in child food.

The Cochrane Collaboration found that "a modest and long term reduction in population salt intake [...] would result in a lower population blood pressure, and a reduction in strokes, heart attacks and heart failure. Furthermore, our study is consistent with the fact that the lower the salt intake, the lower the blood pressure." However, salt consumption is not linked to asthma.

The risk for disease due to insufficient or excessive salt intake varies because of biochemical individuality. Some have asserted that while the risks of consuming too much salt are real, the risks have been exaggerated for most people, or that the studies done on the consumption of salt can be interpreted in many different ways.

Some isolated cultures, such as the Yanomami in South America, have been found to consume little salt, possibly an adaptation originated in the predominantly vegetarian diet of human primate ancestors. However, the low salt diets of the Yanomamo Indians does not result in their low blood pressure, this has been attributed to their lack of a D/D genotype.

Recommended Intake

In the United Kingdom the Scientific Advisory Committee on Nutrition (SACN) recommended in 2003 that, for a typical adult, the Reference Nutrient Intake (RNI) is 4 g salt per day (1.6 g or 70 mmol sodium). However, average adult intake is two and a half times the Reference Nutrient Intake for sodium. SACN states, "The target salt intakes set for adults and children do not represent ideal or optimum consumption levels, but achievable population goals." The Food Safety Authority of Ireland endorses the UK targets. Health Canada recommends an Adequate Intake (AI) and an Upper Limit (UL) in terms of *sodium*, as does The Auckland District Health Board in New

Zealand. Health Canada recommends an AI of 1200–1500 mg and an UL of 2200–2300 mg per day for persons aged 9 years or more.

The NHMRC in Australia was not able to define a recommended dietary intake (RDI). It defines an Adequate Intake (AI) for adults of 460–920 mg/day and an Upper Level of intake (UL) of 2300 mg/day.

In the United States, the Food and Drug Administration itself does not make a recommendation, but refers readers to *Dietary Guidelines for Americans 2005*. These suggest that US citizens should consume less than 2,300 mg of sodium (= 2.3 g sodium = 5.8 g salt) per day.

Meta-analysis in 2009 found that the sodium consumption of 19,151 individuals from 33 countries fit into the narrow range of 2,700 to 4,900 mg/day. The small range across many cultures, together with animal studies, suggest that sodium intake is tightly controlled by feedback loops in the body, making recommendations to reduce sodium consumption below 2,700 mg/day potentially futile.

Labelling

UK: The Food Standards Agency defines the level of salt in foods as follows: "High is more than 1.5 g salt per 100 g (or 0.6 g sodium). Low is 0.3 g salt or less per 100 g (or 0.1 g sodium). If the amount of salt per 100 g is in between these figures, then that is a medium level of salt." In the UK, foods produced by some supermarkets and manufacturers have 'traffic light' colours on the front of the pack: Red (High), Amber (Medium), or Green (Low).

USA: The FDA *Food Labelling Guide* stipulates whether a food can be labelled as "free", "low", or "reduced/less" in respect of sodium. When other health claims are made about a food (e.g. low in fat, calories, etc.), a disclosure statement is required if the food exceeds 480 mg of sodium per 'serving.' Normal salt itself contains 40 g of sodium per 100 g of salt.

Campaigns

In 2004, Britain's Food Standards Agency started a public health campaign called "Salt – Watch it", which recommends no more than 6g of salt per day; it features a character called Sid the Slug and was criticised by the Salt Manufacturers Association (SMA). The Advertising Standards Authority did not uphold the SMA complaint in its adjudication. In March 2007, the FSA launched the third phase of their campaign with the slogan "Salt. Is your food full of it?" fronted by comedienne Jenny Eclair. The University of Tasmania's Menzies Research Institute maintains a website to educate people about the problems of a salt-laden diet.

Consensus Action on Salt and Health (CASH) established in 1996, actively campaigns to raise awareness of the harmful health effects of salt. The 2008 focus includes raising awareness of high levels of salt hidden in sweet foods and marketed towards children.

Taxation of sodium has been proposed as a method of decreasing sodium intake and thereby improving health in countries like the United States where typical salt consumption is high.

The Salt Institute, a salt industry body, is active in promoting the use of salt, and questioning or opposing restrictions on salt intake.

Additives

Iodized salt (BrE: *iodised salt*) is table salt mixed with a minute amount of potassium iodide, sodium iodide, or sodium iodate. Iodized salt is used to help reduce the incidence of iodine deficiency in humans. Iodine deficiency commonly leads to thyroid gland problems, specifically endemic goiter, a disease characterized by a swelling of the thyroid gland, usually resulting in a bulbous protrusion on the neck.

While only tiny quantities of iodine are required in the diet to prevent goiter, the United States Food and Drug Administration recommends [21 CFR 101.9 (c)(8)(iv)] 150 micrograms of iodine per day for both men and women.

Iodized table salt has significantly reduced disorders of iodine deficiency in countries where it is used. Iodine is important to prevent the insufficient production of thyroid hormones (hypothyroidism), which can cause goitre, cretinism in children, and myxedema in adults.

Table salt is mainly employed in cooking and as a table condiment. The amount of iodine and the specific iodine compound added to salt varies from country to country. In the United States, iodized salt contains 46–77 ppm (parts per million), while in the UK the iodine content of iodized salt is recommended to be 10–22 ppm. Today, iodized salt is more common in the United States, Australia and New Zealand than in the United Kingdom.

In some European countries where drinking water fluoridation is not practiced, fluoridated table salt is available. In France, 35% of sold table salt contains either sodium fluoride or potassium fluoride.

Another additive, especially important for pregnant women, is folic acid (vitamin B_9), which gives the table salt a yellow colour.

In Canada, at least one brand (Windsor salt) contains invert sugar.

Sodium ferrocyanide, also known as yellow prussiate of soda, is sometimes added to salt as an anticaking agent. The additive is considered safe for human consumption.

Salt Substitutes

Salt intake can be reduced by simply reducing the quantity of salty foods in a diet, without recourse to salt substitutes. Salt substitutes have a taste similar to table salt and contain mostly potassium chloride, which will increase potassium intake. Excess potassium intake can cause hyperkalemia. Various diseases and medications may decrease the body's excretion of potassium, thereby increasing the risk of hyperkalemia. Those who have kidney failure, heart failure or diabetes should seek medical advice before using a salt substitute. One manufacturer, LoSalt, has issued an advisory statement that those taking the following prescription drugs should not use a salt substitute: amiloride, triamterene, Dytac, spironolactone (Aldactone), and eplerenone (Inspra).

Sodium-Metabolism

Digestive Absorption

Sodium intake from food, primarily in the form of chloride, is 1 to 4 g per day. The digestive absorption of sodium is very fast and almost complete.

Tissue Distribution

In Blood

Almost the total amount of sodium in blood is in the plasma, there is very little in blood cells. Its normal plasma concentration is 140 mmol/L, corresponding to 3.2 g per liter; that of the extravascular fluid is approximately the same.

Hyponatremia, characterized by a plasma concentration lower than 136 mmol/L can result from either a decrease of sodium with normal aqueous volume, or from an expansion of the aqueous volume and dilution of sodium. This last type known as of dilution hyponatremia is often drug-induced, it is observed especially with antidepressants.

Hypernatremia, characterized by a sodium concentration higher than 144 mmol/L, can result from an increase of sodium or a decrease of the aqueous volume.

In Tissues

The cells contain little sodium whereas they are rich in potassium. Bones contain a large amount of sodium: more than 40% of total sodium, that is to say approximately 40 g, is present in bones, in non exchangeable and exchangeable forms.

Elimination

The elimination of sodium is primarily urinary. Its digestive elimination is very low because, although present in exocrine digestive secretions, it can be reabsorbed. Sweat, tears contain sodium but play a negligible part in its elimination.

In the kidney, after glomerular filtration, sodium is actively reabsorbed at the level of the proximal convoluted tubule, the ascending limb of Henle's loop and finally the distal convoluted tubule and collecting duct where its reabsorption is controlled by aldosterone. The regulation of sodium excretion is carried out by the glomerulotubular balance whose mechanism is poorly understood.

Aldosterone decreases the urinary excretion of sodium, whereas atrial natriuretic peptide increases it.

Disorders of Sodium Metabolism

The metabolism of sodium is closely linked to that of water. One can observe hyponatremia and hypernatremia. It is not necessarily a question of lack or excess of sodium, but often the consequences of dissociated variations of water and sodium, a water excess, for example, causes a hyponatremia of dilution.

Hyponatremia can be observed in patients with a mediastinal tumour or a neurological disease or more frequently in patients treated with certain drugs (antidepressants, carbamazepine, neuroleptic agents.

A sufficient intake of water or electrolytes by oral route each time it is possible avoids and corrects the anomalies, while reducing the risk of overloading linked to parenteral administration. However, when the intake by oral route is impossible or when there is an important imbalance, they should be administered by intravenous route.

Precautions

Extreme care is required in handling elemental/metallic sodium. Sodium is potentially explosive in water (depending on quantity), and it is rapidly converted to sodium hydroxide on contact with moisture and sodium hydroxide is a corrosive substance. The powdered form may combust spontaneously in air or oxygen. Sodium must be stored either in an inert (oxygen and moisture free) atmosphere (such as nitrogen or argon), or under a liquid hydrocarbon such as mineral oil or kerosene.

The reaction of sodium and water is a familiar one in chemistry labs, and is reasonably safe if amounts of sodium smaller than a pencil eraser are used and the reaction is done behind a plastic shield by people wearing eye protection. However, the sodium-water reaction does not scale up well, and is treacherous when larger amounts of sodium are used. Larger pieces of sodium melt under the heat of the reaction, and the molten ball of metal is buoyed up by hydrogen and may appear to be stably reacting with water, until splashing covers more of the reaction mass, causing thermal runaway and an explosion which scatters molten sodium, lye solution, and sometimes flame. (18.5 g explosion) This behaviour is unpredictable, and among the alkali metals it is usually sodium which invites this surprise phenomenon, because lithium is not reactive enough to do it, and potassium is so reactive that chemistry students are not tempted to try the reaction with larger potassium pieces.

Sodium is much more reactive than magnesium; a reactivity which can be further enhanced due to sodium's much lower melting point. When sodium catches fire in air (as opposed to just the hydrogen gas generated from water by means of its reaction with sodium) it more easily produces temperatures high enough to melt the sodium, exposing more of its surface to the air and spreading the fire.

Few common fire extinguishers work on sodium fires. Water, of course, exacerbates sodium fires, as do water-based foams. CO_2 and Halon are often ineffective on sodium fires, which reignite when the extinguisher dissipates. Among the very few materials effective on a sodium fire are Pyromet and Met-L-X. Pyromet is a $NaCl/(NH_4)_2HPO_4$ mix, with flow/anti-clump agents. It smothers the fire, drains away heat, and melts to form an impermeable crust. This is the standard dry-powder canister fire extinguisher for all classes of fires. Met-L-X is mostly sodium chloride, NaCl, with approximately 5% Saran plastic as a crust-former, and flow/anti-clumping agents. It is most commonly hand-applied, with a scoop. Other extreme fire extinguishing materials include Lith+, a graphite based dry powder with an organophosphate flame retardant; and Na+, a Na_2CO_3-based material. Alternatively, plain dry sand can effectively slow down the oxygen and humidity flow to the sodium.

Because of the reaction scale problems, disposing of large quantities of sodium (more than 10 to 100 grams) must be done through a licensed hazardous materials disposer. Smaller quantities may be broken up and neutralized carefully with ethanol (which has a much slower reaction than water), or even methanol (where the reaction is more rapid than ethanol's but still less than in water), but care should nevertheless be taken, as the caustic products from the ethanol or methanol reaction are just as hazardous to eyes and skin as those from water. After the alcohol reaction appears complete, and all pieces of reaction debris have been broken up or dissolved, a mixture of alcohol and water, then pure water, may then be carefully used for a final cleaning. This should be allowed to stand a few minutes until the reaction products are diluted more thoroughly and flushed down the drain. The purpose of the final water soaking and washing of any reaction mass or container which may contain sodium, is to ensure that alcohol does not carry unreacted sodium into the sink trap, where a water reaction may generate hydrogen in the trap space which can then be potentially ignited, causing a confined sink trap explosion.

Chapter 7

Homocysteine an Amino Acid

Homocysteine is an amino acid with the formula HSCH$_2$CH$_2$CH(NH$_2$)CO$_2$H. It is a homologue of the amino acid cysteine, differing by an additional methylene (-CH$_2$-) group. It is biosynthesized from methionine by the removal of its terminal C methyl group. Homocysteine can be recycled into methionine or converted into cysteine with the aid of B-vitamins.

While detection of high levels of homocysteine has been linked to cardiovascular disease, lowering homocysteine levels may not improve outcomes.

Homocysteine exists at neutral pH values as a zwitterion.

Betatine form of (S)-homocysteine (left) and (R)-homocysteine (right).

Biosynthesis and Biochemical Roles

Homocysteine is not obtained from the diet. Instead, it is biosynthesized from methionine via a multi-step process. First, methionine receives an adenosine group from ATP, a reaction catalysed by S-adenosyl-methionine synthetase, to give S-adenosyl methionine (SAM). SAM then transfers the methyl group to an acceptor molecule, (i.e., norepinephrine as an acceptor during epinephrine synthesis, DNA methyltransferase as an intermediate acceptor in the process of DNA methylation). The adenosine is then hydrolysed to yield L-homocysteine. L-Homocysteine has two primary fates: conversion via tetrahydrofolate (THF) back into L-methionine or conversion to L-cysteine.

Biosynthesis of Cysteine

Mammals biosynthesize the amino acid cysteine via homocysteine. Cystathionine β-synthase catalyses the condensation of homocysteine and serine to give cystathionine. This reaction uses pyridoxine (vitamin B$_6$) as a cofactor. Cystathionine β-lyase then converts this double amino acid to cysteine, ammonia, and α-ketobutyrate. Bacteria and plants rely on a different pathway to produce cysteine, relying on O-acetylserine.

Methionine Salvage

Homocysteine can be recycled into methionine. This process uses N5-methyl tetrahydrofolate as the methyl donor and cobalamin (vitamin B_{12})-related enzymes. More detail on those enzymes: Tetrahydrofolate-methyltransferase

Other Reactions of Biochemical Significance

Homocysteine can cyclize to give homocysteine thiolactone, a five-membered heterocycle. Because of this "self-looping" reaction, homocysteine-containing peptides tend to cleave themselves.

Influence, Proposed and Verified of Homocysteine on Human Health

Elevated Homocysteine

Deficiencies of the vitamins folic acid (B_9), pyridoxine (B_6), or B_{12} (cyanocobalamin) can lead to high homocysteine levels. Supplementation with pyridoxine, folic acid, B_{12} or trimethylglycine (betaine) reduces the concentration of homocysteine in the bloodstream.

Increased levels of homocysteine are linked to high concentrations of endothelial asymmetric dimethylarginine. Recent research suggests that intense, long duration exercise raises plasma homocysteine levels, perhaps by increasing the load on methionine metabolism.

Elevations of homocysteine also occur in the rare hereditary disease homocystinuria and in the methylene-tetrahydrofolate-reductase polymorphism genetic traits. The latter is quite common (about 10% of the world population) and it is linked to an increased incidence of thrombosis and cardiovascular disease, which occurs more often in people with above minimal levels of homocysteine (about 6 μmol/L).

These individuals require adequate dietary riboflavin in order for homocysteine levels to remain normal. Common levels in Western populations are 10 to 12 and levels of 20 μmol/L are found in populations with low B-vitamin intakes (e.g., New Delhi) or in the older elderly (e.g., Rotterdam, Framingham). Women have 10-15% less homocysteine during their reproductive decades than men, which may help explain the fact they suffer myocardial infarction (heart attacks) on average 10 to 15 years later than men. However, this phenomenon is more readily explained by higher levels of estrogen, which exerts a cardioprotective effect.

Blood reference ranges for homocysteine:

Sex	Age	Lower limit	Upper limit	Unit	Elevated	Therapeutic target
Female	12–19 years	3.3	7.2	µmol/L	> 10.4	< 6.3 (0.85 mg/L)
	>60 years	4.9	11.6	µmol/L		
Male	12–19 years	4.3	9.9	µmol/L	> 11.4	
	>60 years	5.9	15.3	µmol/L		

Cardiovascular Risks and related Medical Studies

A high level of blood serum homocysteine "homocysteinemia" is a powerful risk factor for cardiovascular disease. Unfortunately, one study which attempted to decrease the risk by lowering homocysteine was not fruitful. This study was conducted on nearly 5000 Norwegian heart attack survivors who already had severe, late-stage heart disease. No study has yet been conducted in a preventive capacity on subjects who are in a relatively good state of health.

Studies reported in 2006 have shown that giving vitamins [folic acid, B_6 and B_{12}] to reduce homocysteine levels may not quickly offer benefit, however a significant 25% reduction in stroke was found in the HOPE-2 study even in patients mostly with existing serious arterial decline although the overall death rate was not significantly changed by the intervention in the trial. Clearly, reducing homocysteine does not quickly repair existing structural damage of the artery architecture. However, the science is strongly supporting the biochemistry that homocysteine degrades and inhibits the formation of the three main structural components of the artery, collagen, elastin and the proteoglycans. Homocysteine permanently degrades cysteine disulfide bridges and lysine amino acid residues in proteins, gradually affecting function and structure. Simply put, homocysteine is a 'corrosive' of long-living proteins, i.e., collagen or elastin, or life-long proteins, i.e., fibrillin. These long-term effects are difficult to establish in clinical trials focusing on groups with existing artery decline. The main role of reducing homocysteine is possibly in 'prevention' but studies in patients with pre-existing conditions found no significant benefit nor damage.

Hypotheses have been offered to address the failure of homocysteine-lowering therapies to reduce cardiovascular event frequency. One suggestion is that folic acid may directly cause an increased build-up of arterial plaque, independent of its homocysteine-lowering effects. Alternatively, folic acid and vitamin B_{12} may cause an overall change in gene methylation levels in vascular cells, which may also promote plaque growth. Finally, altering methlyation activity in cells might increase methylation of l-arginine to asymmetric dimethylarginine which can increase the risk of vascular disease. Thus alternative homocysteine-lowering therapies may yet be developed which show greater effects on development and progression of cardiovascular disease.

The VITATOPS trial (results presented in May 2010 by the lead investigator, Dr Graeme J Hankey of Royal Perth Hospital, Australia at the European Stroke Conference 2010, in Barcelona, Spain) has concluded that B-vitamin supplements, within 2 years, do not seem to significantly reduce subsequent stroke, MI, or vascular death in patients with a history of recent stroke and ischemic attack, despite lowering of homocysteine levels.

Bone Weakness and Breaks

Elevated levels of homocysteine have been linked to increased fractures in elderly persons. The high level of homocysteine will auto-oxidize and react with reactive oxygen intermediates and damage endothelial cells and has a higher risk to form a thrombus. Homocysteine does not affect bone density. Instead, it appears that homocysteine affects collagen by interfering with the cross-linking between the collagen fibers and the tissues they reinforce. Whereas the HOPE-2 trial showed a reduction in stroke incidence, in those with stroke there is a high rate of hip fractures in the affected side. A trial with 2 homocysteine-lowering vitamins (folate and B_{12}) in people with prior stroke, there was an 80% reduction in fractures, mainly hip, after 2 years. Interestingly, also here, bone density (and the number of falls) were identical in the vitamin and the placebo groups. Vitamin supplements counter the deleterious effects of homocysteine on collagen. As they inefficiently absorb B_{12} from food, elderly persons may benefit from taking higher doses orally such as 100 mcg/day (found in some multivitamins) or by intramuscular injection.

Cofactors or Coenzymes: Nature's Special Reagents

A cofactor is a non-protein chemical compound that is bound to a protein and is required for the protein's biological activity. These proteins are commonly enzymes, and cofactors can be considered "helper molecules" that assist in biochemical transformations. Cofactors are either organic or inorganic. They can also be classified depending on how tightly they bind to an enzyme, with loosely-bound cofactors termed coenzymes and tightly-bound cofactors termed prosthetic groups. Some sources also limit the use of the term "cofactor" to inorganic substances. An inactive enzyme, without the cofactor is called an apoenzyme, while the complete enzyme with cofactor is the holoenzyme.

Some enzymes or enzyme complexes require several cofactors, for example the multienzyme complex pyruvate dehydrogenase. This enzyme complex at the junction of glycolysis and the citric acid cycle requires five organic cofactors and one metal ion: loosely bound thiamine pyrophosphate (TPP), covalently bound lipoamide and flavin adenine dinucleotide (FAD), and the cosubstrates nicotinamide adenine dinucleotide (NAD) and coenzyme A (CoA) and a metal ion (Mg^2). Organic cofactors are often vitamins or are made from vitamins. Many contain the nucleotide adenosine monophosphate (AMP) as part of their structures, such as ATP, coenzyme A, FAD and NAD^+. This common structure may reflect a common evolutionary origin as part of ribozymes in an ancient RNA world. It has been suggested that the AMP part of the molecule can be considered a kind of "handle" by which the enzyme can "grasp" the coenzyme to switch it between different catalytic centres.

Classification

Cofactors can be divided into two broad groups: organic cofactors, such as flavin or heme, and inorganic cofactors: such as the metal ions Mg^{2+}, Cu^+, Mn^{2+} or iron-sulfur clusters.

Organic cofactors are sometimes further divided into *coenzymes* and *prosthetic groups*. The term coenzyme refers specifically to enzymes and as such to the functional properties of a protein. On the other hand "prosthetic group", emphasizes the nature of the binding of a cofactor to a protein (tight or covalent) and thus refers to a structural property. Different sources give slightly different definitions of coenzymes, cofactors and prosthetic groups. Some consider tightly-bound organic molecules as prosthetic groups and not as coenzymes, while others define all non-protein organic molecules needed for enzyme activity as coenzymes, and classify those that are tightly bound as coenzyme prosthetic groups. Unsurprisingly, these terms are often used loosely. A 1979 letter in *Trends in Biochemical Sciences* noted the confusion in the literature and the essentially arbitrary distinction made between prosthetic groups and coenzymes and proposed the following scheme. Here, cofactors were defined as an additional substance apart from protein and substrate that is required for enzyme activity and a prosthetic group as a substance that undergoes its whole catalytic cycle attached to a single enzyme molecule. However, the author could not arrive at a single all-encompassing definition of a "coenzyme" and proposed that this term be dropped from use in the literature.

Inorganic

Iron-sulfur Clusters

Iron-sulfur clusters are complexes of iron and sulfur atoms held within proteins by cysteinyl residues. They play both structural and functional roles, including electron transfer, redox sensing, and as structural modules. Iron-sulfur proteins are proteins characterized by the presence of iron-sulfur clusters containing sulfide-linked di-, tri-, and tetrairon centres in variable oxidation states. Iron-sulfur clusters are found in a variety of metalloproteins, such as the ferredoxins, as well as NADH dehydrogenase, hydrogenases, Coenzyme Q- cytochrome c reductase, Succinate-coenzyme Q reductase and nitrogenase. Iron-sulfur clusters are best known for their role in the oxidation-reduction reactions of mitochondrial electron transport. Both Complex I and Complex II of oxidative phosphorylation have multiple Fe-S clusters. They have many other functions including catalysis as illustrated by aconitase, generation of radicals as illustrated by SAM-dependent enzymes, and as sulfur donors in the biosynthesis of lipoic acid and biotin. Additionally some Fe-S proteins regulate gene expression. Fe-S proteins are vulnerable to attack by biogenic nitric oxide.

Metal Ions

Metal ions are common cofactors. The study of these cofactors falls under the area of bioinorganic chemistry. In nutrition, the list of essential trace elements reflects their role as cofactors. In humans this list commonly includes iron, magnesium, manganese, cobalt, copper, zinc, selenium, and molybdenum. Although chromium deficiency causes impaired glucose tolerance, no human enzyme that uses this metal as a cofactor has been identified. Iodine is also an essential trace element, but this element is used as part of the structure of thyroid hormones rather than as an enzyme cofactor. Calcium is another special case, in that it is required as a component of the human diet, and it is needed for the full activity of many enzymes:

such as nitric oxide synthase, protein phosphatases or adenylate kinase, but calcium activates these enzymes in allosteric regulation, often binding to these enzymes in a complex with calmodulin. Calcium is therefore a cell signalling molecule, and not usually considered as a cofactor of the enzymes it regulates.

Other organisms require additional metals as enzyme cofactors, such as vanadium in the nitrogenase of the nitrogen-fixing bacteria of the genus *Azotobacter*, tungsten in the aldehyde ferredoxin oxidoreductase of the thermophilic archaean *Pyrococcus furiosus*, and even cadmium in the carbonic anhydrase from the marine diatom *Thalassiosira weissflogii*.

In many cases, the cofactor includes both an inorganic and organic component. One diverse set of examples are the haem proteins, which consists of a porphyrin ring coordinated to iron.

Ion	Examples of enzymes containing this ion
Cupric	Cytochrome oxidase
Ferrous or Ferric	Catalase
	Cytochrome (via Heme)
	Nitrogenase
	Hydrogenase
Magnesium	Glucose 6-phosphatase
	Hexokinase
	DNA polymerase
Manganese	Arginase
Molybdenum	Nitrate reductase
Nickel	Urease
Selenium	Glutathione peroxidase
Zinc	Alcohol dehydrogenase
	Carbonic anhydrase
	DNA polymerase

Biosynthesis

The biosynthesis of the Fe-S clusters has been well studied. The biogenesis of iron sulfur clusters has been studied most extensively in the bacteria *E. coli* and *A. vinelandii* and yeast *S. cerevisiae*. At least three different biosynthetic systems have been identified so far, namely nif, suf, and isc systems, which were first identified in bacteria.

The nif system is responsible the clusters in the enzyme nitrogenase. The suf and isc systems are more general with the isc-related proteins being present only in the animal kingdom. The yeast isc system is the best described. Several proteins constitute the biosynthetic machinery via the isc pathway. The process occurs in two major steps: (1) the Fe/S cluster is assembled on a scaffold protein followed by (2) transfer of the preformed cluster to the recipient proteins. The first step of this process occurs in the cytoplasm of prokaryotic organisms or in the mitochondria of eukaryotic organisms. In the higher organisms the clusters are therefore transported out of the mitochondrion to be incorporated into the extramitochondrial enzymes. These organisms also possess a set of proteins involved in the Fe/S clusters transport and incorporation processes that are not homologous to proteins found in procaryotic systems.

Synthetic Analogues

Synthetic analogues of the naturally occurring Fe-S clusters were first reported by Holm and coworkers. Treatment of iron salts with a mixture of thiolates and sulfide affords derivatives such as $(Et_4N)_2Fe_4S_4(SCH_2Ph)_4]$.

Organic

Organic cofactors are small organic molecules (typically a molecular mass less than 1000 Da) that can be either loosely or tightly bound to the enzyme and directly participate in the reaction. In the latter case, when it is difficult to remove without denaturing the enzyme, it can be called a prosthetic group. It is important to emphasize that there is no sharp division between loosely and tightly bound cofactors. Indeed, many, such as NAD^+ can be tightly bound in some enzymes, while it is loosely bound in others. Another example is thiamine pyrophosphate (TPP) is tightly bound in transketolase or pyruvate decarboxylase, while it is less tightly bound in pyruvate dehydrogenase. Other coenzymes, flavin adenine dinucleotide (FAD), biotin or lipoamide for instance, are covalently bound. Tightly-bound cofactors are generally regenerated during the same reaction cycle, while loosely-bound cofactors can be regenerated in a subsequent reaction catalysed by a different enzyme. In the latter case, the cofactor can also be considered a substrate or cosubstrate.

Vitamins can serve as precursors to many organic cofactors (e.g. vitamins B_1, B_2, B_6, B_{12}, niacin, folic acid) or as coenzymes themselves (e.g. vitamin C). However, vitamins do have other functions in the body. Many organic cofactors also contain a nucleotide: such as the electron carriers NAD and FAD, or coenzyme A, which carries acyl groups. Most of these cofactors are found in a huge variety of species, and some are universal to all forms of life. An exception to this wide distribution is a group of unique cofactors that evolved in methanogens, which are restricted to this group of archaea.

Cofactors as Metabolic Intermediates

Metabolism involves a vast array of chemical reactions, but most fall under a few basic types of reactions that involve the transfer of functional groups. This common chemistry allows cells to use a small set of metabolic intermediates to carry chemical groups between different reactions. These group-transfer intermediates are the loosely-bound organic cofactors, often called *coenzymes*.

Each class of group-transfer reaction is carried out by a particular cofactor, which is the substrate for a set of enzymes that produce it, and a set of enzymes that consume it. An example of this are the dehydrogenases that use nicotinamide adenine dinucleotide (NAD^+) as a cofactor. Here, hundreds of separate types of enzymes remove electrons from their substrates and reduce NAD^+ to NADH. This reduced cofactor is then a substrate for any of the reductases in the cell that require electrons to reduce their substrates.

These cofactors are therefore continuously recycled as part of metabolism. As an example, the total quantity of ATP in the human body is about 0.1 mole. This ATP is constantly being broken down into ADP, and then converted back into ATP. Thus, at any given time, the total amount of ATP + ADP remains fairly constant. The energy used by human cells requires the hydrolysis of 100 to 150 moles of ATP daily which is around 50 to 75 kg. Typically, a human will use up their body weight of ATP over the course of the day. This means that each ATP molecule is recycled 1000 to 1500 times daily.

Evolution

Organic cofactors, such as ATP and NADH, are present in all known forms of life and form a core part of metabolism. Such universal conservation indicates that these molecules evolved very early in the development of living things. At least some of the current set of cofactors may therefore have been present in the last universal ancestor, which lived about 4 billion years ago.

Organic cofactors may have been present even earlier in the history of life on Earth. Interestingly, the nucleotide adenosine is present in cofactors that catalyse many basic metabolic reactions such as methyl, acyl, and phosphoryl group transfer, as well as redox reactions. This ubiquitous chemical scaffold has therefore been proposed to be a remnant of the RNA world, with early ribozymes evolving to bind a restricted set of nucleotides and related compounds. Adenosine-based cofactors are thought to have acted as interchangeable adaptors that allowed enzymes and ribozymes to bind new cofactors through small modifications in existing adenosine-binding domains, which had originally evolved to bind a different cofactor. This process of adapting a pre-evolved structure for a novel use is referred to as *exaptation*.

Abiogenesis

In natural science, abiogenesis or biopoesis is the study of how life on Earth arose from inanimate matter. Most amino acids, often called "the building blocks of life", can form via natural chemical reactions unrelated to life, as demonstrated in the Miller–Urey experiment and similar experiments, which involved simulating some of the conditions of the early Earth, in a scientific laboratory. In all living things, these amino acids are organized into proteins, and the construction of these proteins is mediated by nucleic acids. Which of these organic molecules first arose and how they formed the first life is the focus of abiogenesis.

In any theory of abiogenesis, two aspects of life have to be accounted for: replication, and metabolism. The question of which came first gave rise to different types of theories. In the beginning, metabolism-first theories (Oparin coacervate) were proposed, and only later thinking gave rise to the modern, replication-first approach.

In modern, still somewhat limited understanding, the first living things on Earth are thought to be single cell prokaryotes (which lack a cell nucleus), perhaps evolved from protobionts (organic molecules surrounded by a membrane-like structure). The oldest ancient fossil microbe-like objects are dated to be 3.5 Ga (billion years old), approximately one billion years after the formation of the Earth itself. By 2.4 Ga, the ratio of stable isotopes of carbon, iron and sulfur shows the action of living things on inorganic minerals and sediments and molecular biomarkers indicate photosynthesis, demonstrating that life on Earth was widespread by this time.

The sequence of chemical events that led to the first nucleic acids is not known. Several hypotheses about early life have been proposed, most notably the iron-sulfur world theory (metabolism without genetics) and the RNA world hypothesis (RNA life-forms).

Conceptual History

Spontaneous Generation

Until the early 19th century, people generally believed in the ongoing spontaneous generation of certain forms of life from non-living matter. This was paired with heterogenesis, the belief that one form of life derives from a different form (*e.g.* bees from flowers). Classical notions of abiogenesis, now more precisely known as *spontaneous generation,* held that certain complex, living organisms are generated by decaying organic substances. According to Aristotle it was a readily observable truth that aphids arise from the dew which falls on plants, flies from putrid matter, mice from dirty hay, crocodiles from rotting logs at the bottom of bodies of water, and so on.

In the 17th century, such assumptions started to be questioned; for example, in 1646, Sir Thomas Browne published his *Pseudodoxia Epidemica* (subtitled *Enquiries into Very many Received Tenets, and Commonly Presumed Truths*), which was an attack on false beliefs and "vulgar errors." His conclusions were not widely accepted. For example, his contemporary, Alexander Ross wrote: "To question this (i.e., spontaneous generation) is to question reason, sense and experience. If he doubts of this let him go to Egypt, and there he will find the fields swarming with mice, begot of the mud of Nylus, to the great calamity of the inhabitants."

In 1665, Robert Hooke published the first drawings of a microorganism. Hooke was followed in 1676 by Anton van Leeuwenhoek, who drew and described microorganisms that are now thought to have been protozoa and bacteria. Many felt the existence of microorganisms was evidence in support of spontaneous generation, since microorganisms seemed too simplistic for sexual reproduction, and asexual reproduction through cell division had not yet been observed.

The first solid evidence against spontaneous generation came in 1668 from Francesco Redi, who proved that no maggots appeared in meat when flies were prevented from laying eggs. It was gradually shown that, at least in the case of all the higher and readily visible organisms, the previous sentiment regarding spontaneous generation was false. The alternative seemed to be biogenesis: that every living thing came from a pre-existing living thing (*omne vivum ex ovo*, Latin for "every living thing from an egg").

In 1768, Lazzaro Spallanzani demonstrated that microbes were present in the air, and could be killed by boiling. In 1861, Louis

Pasteur performed a series of experiments which demonstrated that organisms such as bacteria and fungi do not spontaneously appear in sterile, nutrient-rich media.

Pasteur and Darwin

By the middle of the 19th century, the theory of biogenesis had accumulated so much evidential support, due to the work of Louis Pasteur and others, that the alternative theory of spontaneous generation had been effectively disproven. Pasteur himself remarked, after a definitive finding in 1864, "Never will the doctrine of spontaneous generation recover from the mortal blow struck by this simple experiment." The collapse of spontaneous generation, however, left a vacuum of scientific thought on the question of how life *had* first arisen.

In a letter to Joseph Dalton Hooker on February 1, 1871, Charles Darwin addressed the question, suggesting that the original spark of life may have begun in a "warm little pond, with all sorts of ammonia and phosphoric salts, lights, heat, electricity, etc. present, so that a protein compound was chemically formed ready to undergo still more complex changes". He went on to explain that "at the present day such matter would be instantly devoured or absorbed, which would not have been the case before living creatures were formed." In other words, the presence of life itself makes the search for the origin of life dependent on the sterile conditions of the laboratory.

"Primordial Soup" Theory

No new notable research or theory on the subject appeared until 1924, when Alexander Oparin reasoned that atmospheric oxygen prevents the synthesis of certain organic compounds that are necessary building blocks for the evolution of life. In his *The Origin of Life*, Oparin proposed that the "spontaneous generation of life" that had been attacked by Louis Pasteur, did in fact occur once, but was now impossible because the conditions found in the early earth had changed, and the presence of living organisms would immediately consume any spontaneously generated organism. Oparin argued that a "primeval soup" of organic molecules could be created in an oxygen-less atmosphere through the action of sunlight. These would combine in ever-more complex fashions until they formed coacervate droplets. These droplets would "grow" by fusion with other droplets, and "reproduce" through fission into daughter droplets, and so have a primitive metabolism in which those factors which promote "cell integrity" survive, and those that do not become extinct. Many modern theories of the origin of life still take Oparin's ideas as a starting point. Around the same time, J. B. S. Haldane suggested that the Earth's pre-biotic oceans–very different from their modern counterparts–would have formed a "hot dilute soup" in which organic compounds could have formed. This idea was called *biopoiesis* or *biopoesis*, the process of living matter evolving from self-replicating but nonliving molecules.

Early Conditions

Morse and MacKenzie have suggested that oceans may have appeared first in the Hadean eon, as soon as two hundred million years (200 Ma) after the Earth was formed, in a hot 100 °C (212 °F) reducing environment, and that the pH of about 5.8 rose rapidly towards neutral. This has been supported by Wilde who has pushed the date of the zircon crystals found in the metamorphosed quartzite of Mount Narryer in Western Australia, previously thought to be 4.1–4.2 Ga, to 4.404 Ga. This means that oceans and continental crust existed within 150 Ma of Earth's formation.

Despite this, the Hadean environment was one highly hazardous to life. Frequent collisions with large objects, up to 500 kilometres (310 mi) in diameter, would have been sufficient to vaporise the ocean within a few months of impact, with hot steam mixed with rock vapour leading to high altitude clouds completely covering the planet. After a few months the height of these clouds would have begun to decrease but the cloud base would still have been elevated for about the next thousand years. After that, it would have begun to rain at low altitude. For another two thousand years rains would slowly have drawn down the height of the clouds, returning the oceans to their original depth only 3,000 years after the impact event.

Between 3.8 and 4.1 Ga, changes in the orbits of the gaseous giant planets may have caused a late heavy bombardment that pockmarked the moon and other inner planets (Mercury, Mars, and presumably Earth and Venus). This would likely have sterilized the planet had life appeared before that time.

By examining the time interval between such devastating environmental events, the time interval when life might first have come into existence can be found for different early environments. The study by Maher and Stevenson shows that if the deep marine hydrothermal setting provides a suitable site for the origin of life, abiogenesis could have happened as early as 4.0 to 4.2 Ga, whereas if it occurred at the surface of the earth abiogenesis could only have occurred between 3.7 and 4.0 Ga.

Other research suggests a colder start to life. Work by Leslie Orgel and colleagues on the synthesis of purines has shown that freezing temperatures are advantageous, due to the concentrating effect for key precursors such as hydrogen cyanide. Research by Stanley Miller and colleagues suggested that while adenine and guanine require freezing conditions for synthesis, cytosine and uracil may require boiling temperatures. Based on this research, Miller suggested a beginning of life involving freezing conditions and exploding meteorites. An article in Discover Magazine points to research by the Miller group indicating the formation of seven different amino acids and 11 types of nucleobases in ice when ammonia and cyanide were left in a freezer from 1972–1997. This article also describes research by Christof Biebricher showing the formation of RNA molecules 400 bases long under freezing conditions using an RNA template, a single-strand chain of RNA that guides the formation of a new strand of RNA. As that new RNA strand grows, it adheres to the template. The explanation given for the unusual speed of these reactions at such a low temperature is eutectic freezing. As an ice crystal forms, it stays pure: only molecules of water join the growing crystal, while impurities like salt or cyanide are excluded. These impurities become crowded in microscopic pockets of liquid within the ice, and this crowding causes the molecules to collide more often.

Evidence of the early appearance of life comes from the Isua supercrustal belt in Western Greenland and from similar formations in the nearby Akilia Islands. Carbon entering into rock formations has a ratio of Carbon-13 (^{13}C) to Carbon-12 (^{12}C) of about "5.5 (in units of $\delta^{13}C$), where because of a preferential biotic uptake of ^{12}C, biomass has a $\delta^{13}C$ of between "20 and "30. These isotopic fingerprints are preserved in the sediments, and Mojzis has used this technique to suggest that life existed on the planet already by 3.85 billion years ago. Lazcano and Miller (1994) suggest that the rapidity of the evolution of life is dictated by the rate of recirculating water through mid-ocean submarine vents. Complete recirculation takes 10 million years, thus any organic compounds produced by then would be altered or destroyed by temperatures exceeding 300 °C (572 °F). They estimate that the development of a 100 kilobase genome of a DNA/protein primitive heterotroph into a 7000 gene filamentous cyanobacterium would have required only 7 Ma.

Current Models

There is no truly "standard model" of the origin of life. Most currently accepted models draw at least some elements from the framework laid out by the Oparin-Haldane hypothesis. Under that umbrella, however, are a wide array of disparate discoveries and conjectures such as the following, listed in a rough order of postulated emergence:

1. Some theorists suggest that the atmosphere of the early Earth may have been chemically reducing in nature, composed primarily of methane (CH_4), ammonia (NH_3), water (H_2O), hydrogen sulfide (H_2S), carbon dioxide (CO_2) or carbon monoxide (CO), and phosphate (PO_4^{3-}), with molecular oxygen (O_2) and ozone (O_3) either rare or absent.
2. In such a reducing atmosphere, electrical activity can catalyse the creation of certain basic small molecules (monomers) of life, such as amino acids. This was demonstrated in the Miller–Urey experiment by Stanley L. Miller and Harold C. Urey in 1953.
3. Phospholipids (of an appropriate length) can spontaneously form lipid bilayers, a basic component of the cell membrane.
4. A fundamental question is about the nature of the first self-replicating molecule. Since replication is accomplished in modern cells through the cooperative action of proteins and nucleic acids, the major schools of thought about how the process originated can be broadly classified as "proteins first" and "nucleic acids first".
5. The principal thrust of the "nucleic acids first" argument is as follows:
    1. The polymerization of nucleotides into random RNA molecules might have resulted in self-replicating ribozymes (RNA world hypothesis)
    2. Selection pressures for catalytic efficiency and diversity might have resulted in ribozymes which catalyse peptidyl transfer (hence formation of small proteins), since oligopeptides complex with RNA to form better catalysts. The first ribosome might have been created by such a process, resulting in more prevalent protein synthesis.
    3. Synthesized proteins might then outcompete ribozymes in catalytic ability, and therefore become the dominant biopolymer, relegating nucleic acids to their modern use, predominantly as a carrier of genomic information.

As of 2010, no one has yet synthesized a "protocell" using basic components which would have the necessary properties of life (the so-called *"bottom-up-approach"*). Without such a proof-of-principle, explanations have tended to be short on specifics. However, some researchers are working in this field, notably Steen Rasmussen at Los Alamos National Laboratory and Jack Szostak at Harvard University. Others have argued that a *"top-down approach"* is more feasible. One such approach, successfully attempted by Craig Venter and others at The Institute for Genomic Research, involves engineering existing prokaryotic cells with progressively fewer genes, attempting to discern at which point the most minimal requirements for life were reached. The biologist John Desmond Bernal coined the term Biopoesis for this process, and suggested that there were a number of clearly defined "stages" that could be recognised in explaining the origin of life.

- Stage 1: The origin of biological monomers
- Stage 2: The origin of biological polymers
- Stage 3: The evolution from molecules to cell

Bernal suggested that evolution may have commenced early, some time between Stage 1 and 2.

Origin of Organic Molecules

There are two possible sources of organic molecules on the early Earth:

1. Terrestrial origins–organic synthesis driven by impact shocks or by other energy sources (such as ultraviolet light or electrical discharges) (eg.Miller's experiments)
2. Extraterrestrial origins–delivery by objects (e.g. carbonaceous chondrites) or gravitational attraction of organic molecules or primitive life-forms from space.

Recently, estimates of these sources suggest that the heavy bombardment before 3.5 Ga within the early atmosphere made available quantities of organics comparable to those produced by other energy sources.

"Soup" Theory Today: Miller's Experiment and Subsequent Work

Biochemist Robert Shapiro has summarized the "Primordial Soup" theory of Oparin and Haldane in its "mature form" as follows:

1. The early Earth had a chemically reducing atmosphere.
2. This atmosphere, exposed to energy in various forms, produced simple organic compounds ("monomers").
3. These compounds accumulated in a "soup", which may have been concentrated at various locations (Shorelines, oceanic vents etc.).

4. By further transformation, more complex organic polymers— and ultimately life— developed in the soup.

Regarding the Reducing Atmosphere

Whether the mixture of gases used in the Miller–Urey experiment truly reflects the atmospheric content of early Earth is a controversial topic. Other less reducing gases produce a lower yield and variety. It was once thought that appreciable amounts of molecular oxygen were present in the prebiotic atmosphere, which would have essentially prevented the formation of organic molecules; however, the current scientific consensus is that such was not the case.

Regarding Monomer Formation

One of the most important pieces of experimental support for the "soup" theory came in 1953. A graduate student, Stanley Miller, and his professor, Harold Urey, performed an experiment that demonstrated how organic molecules could have spontaneously formed from inorganic precursors, under conditions like those posited by the Oparin-Haldane Hypothesis. The now-famous "Miller–Urey experiment" used a highly reduced mixture of gases–methane, ammonia and hydrogen–to form basic organic monomers, such as amino acids. This provided direct experimental support for the second point of the "soup" theory, and it is around the remaining two points of the theory that much of the debate now centres.

Apart from the Miller–Urey experiment, the next most important step in research on prebiotic organic synthesis was the demonstration by Joan Oró that the nucleic acid purine base, adenine, was formed by heating aqueous ammonium cyanide solutions. In support of abiogenesis in eutectic ice, more recent work demonstrated the formation of s-triazines (alternative nucleobases), pyrimidines (including cytosine and uracil), and adenine from urea solutions subjected to freeze-thaw cycles under a reductive atmosphere (with spark discharges as an energy source).

Regarding Monomer Accumulation

The "soup" theory relies on the assumption proposed by Darwin that in an environment with no pre-existing life, organic molecules may have accumulated and provided an environment for chemical evolution.

Regarding Further Transformation

The spontaneous formation of complex polymers from abiotically generated monomers under the conditions posited by the "soup" theory is not at all a straightforward process. Besides the necessary basic organic monomers, compounds that would have prohibited the formation of polymers were formed in high concentration during the Miller–Urey and Oró experiments. The Miller experiment, for example, produces many substances that would undergo cross-reactions with the amino acids or terminate the peptide chain.

More fundamentally, it can be argued that the most crucial challenge unanswered by this theory is how the relatively simple organic building blocks polymerise and form more complex structures, interacting in consistent ways to form a protocell. For example, in an aqueous environment hydrolysis of oligomers/polymers into their constituent monomers would be favoured over the condensation of individual monomers into polymers.

The Deep Sea Vent Theory

The deep sea vent, or hydrothermal vent, theory for the origin of life on Earth posits that life may have begun at submarine hydrothermal vents, where hydrogen-rich fluids emerge from below the sea floor and interface with carbon dioxide-rich ocean water. Sustained chemical energy in such systems is derived from redox reactions, in which electron donors, such as molecular hydrogen, react with electron acceptors, such as carbon dioxide.

Fox's Experiments

In the 1950s and 1960s, Sidney W. Fox studied the spontaneous formation of peptide structures under conditions that might plausibly have existed early in Earth's history. He demonstrated that amino acids could spontaneously form small peptides. These amino acids and small peptides could be encouraged to form closed spherical membranes, called protenoid microspheres, which show many of the basic characteristics of 'life'.

Eigen's Hypothesis

In the early 1970s the problem of the origin of life was approached by Manfred Eigen and Peter Schuster of the Max Planck Institute for Biophysical Chemistry. They examined the transient stages between the molecular chaos and a self-replicating hypercycle in a prebiotic soup.

In a hypercycle, the information storing system (possibly RNA) produces an enzyme, which catalyzes the formation of another information system, in sequence until the product of the last aids in the formation of the first information system. Mathematically treated, hypercycles could create quasispecies, which through natural selection entered into a form of Darwinian evolution. A boost to hypercycle theory was the discovery that RNA, in certain circumstances, forms itself into ribozymes, capable of catalysing their own chemical reactions. However, these reactions are limited to self-excisions (in which a longer RNA molecule becomes shorter), and much rarer small additions that are incapable of coding for any useful protein. The hypercycle theory is further degraded since the hypothetical RNA would require the existence of complex biochemicals such as nucleotides which are not formed under the conditions proposed by the Miller–Urey experiment.

Hoffmann's Contributions

Geoffrey W. Hoffmann, a student of Eigen, contributed to the concept of life involving both replication and metabolism emerging from catalytic noise. His contributions included showing that an early sloppy translation machinery can be stable against an error catastrophe of the type that had been envisaged as problematical by Leslie Orgel ("Orgel's paradox") and calculations regarding the occurrence of a set of required catalytic activities together with the exclusion of catalytic activities that would be disruptive. This is called the stochastic theory of the origin of life.

Wächtershäuser's Hypothesis

Another possible answer to this polymerization conundrum was provided in 1980s by the German chemist Günter Wächtershäuser, in his iron-sulfur world theory. In this theory, he postulated the evolution of (bio)chemical pathways as fundamentals of the evolution of life. Moreover, he presented a consistent system of tracing today's biochemistry back to ancestral reactions that provide alternative pathways to the synthesis of organic building blocks from simple gaseous compounds.

In contrast to the classical Miller experiments, which depend on external sources of energy (such as simulated lightning or UV irradiation), "Wächtershäuser systems" come with a built-in source of energy, sulfides of iron and other minerals (e.g. pyrite). The energy released from redox reactions of these metal sulfides is not only available for the synthesis of organic molecules, but also for the formation of oligomers and polymers. It is therefore hypothesized that such systems may be able to evolve into autocatalytic sets of self-replicating, metabolically active entities that would predate the life forms known today.

The experiment produced a relatively small yield of dipeptides (0.4% to 12.4%) and a smaller yield of tripeptides (0.10%) but the authors also noted that: "under these same conditions dipeptides hydrolysed rapidly."

Radioactive beach Hypothesis

Zachary Adam at the University of Washington, Seattle, claims that stronger tidal processes from a much closer moon may have concentrated grains of uranium and other radioactive elements at the high water mark on primordial beaches where they may have been responsible for generating life's building blocks. According to computer models reported in *Astrobiology*, a deposit of such radioactive materials could show the same self-sustaining nuclear reaction as that found in the Oklo uranium ore seam in Gabon. Such radioactive beach sand provides sufficient energy to generate organic molecules, such as amino acids and sugars from acetonitrile in water. Radioactive monazite also releases soluble phosphate into regions between sand-grains, making it biologically "accessible". Thus amino acids, sugars and soluble phosphates can all be simultaneously produced, according to Adam. Radioactive actinides, then in greater concentrations, could have formed part of organo-metallic complexes. These complexes could have been important early catalysts to living processes.

John Parnell of the University of Aberdeen suggests that such a process could provide part of the "crucible of life" on any early wet rocky planet, so long as the planet is large enough to have generated a system of plate tectonics which brings radioactive minerals to the surface. As the early Earth is believed to have had many smaller "platelets" it would provide a suitable environment for such processes.

Thermodynamic Origin of Life: Ultraviolet and Temperature-Assisted Replication (UVTAR) Model

Karo Michaelian of the National Autonomous University of Mexico (UNAM) points out that any model for the origin of life must take into account the fact that life is an irreversible thermodynamic process which arises and persists to produce entropy. Entropy production is not incidental to the process of life, but rather the fundamental reason for its existence. Present day life augments the entropy production of Earth by catalysing the water cycle through evapotranspiration. Michaelian argues that if the thermodynamic function of life today is to produce entropy through coupling with the water cycle, then this probably was its function at its very beginnings. It turns out that both RNA and DNA when in water solution are very strong absorbers and extremely rapid dissipaters of ultraviolet light within the 200 nm - 300 nm wavelength range, just that high energy part of the sun's spectrum that could have penetrated the dense prebiotic atmosphere. Cnossen et al. have shown that the amount of UV light reaching the Earth's surface in the Archean could have been up to 31 orders of magnitude larger than it is today at 260 nm where RNA and DNA absorb most strongly. Absorption and dissipation of UV light by these organic molecules at the Archean ocean surface would have increased significantly the temperature of the surface skin layer leading to enhanced evaporation and thus augmenting the primitive water cycle. Since absorption and dissipation of high energy photons is an entropy producing process, Michaelian argues that non-equilbrium abiogenic synthesis of RNA and DNA utilizing UV light would have been thermodynamically favoured.

A simple mechanism to explain the replication of RNA and DNA without the use of enzymes can also be given within the same thermodynamic framework by assuming that life arose when the temperature of the primitive seas had cooled to somewhat below the denaturing temperature of RNA or DNA (based on the ratio of $^{17}O/^{16}O$ found in cherts of the Barberton greenstone belt of South Africa of about 3.5 to 3.2 Ga., surface temperatures are predicted to have been around 70±15 °C, similar to RNA or DNA denaturing temperatures). During the night, the surface water temperature would be below the denaturing temperature and single strand RNA/DNA could act as a template for the formation of double strand RNA/DNA. During the daylight hours, RNA and DNA would absorb UV light and convert this directly to heating of the ocean surface, raising the local temperature enough to allow for denaturing of RNA and DNA. The copying process would be repeated during the cool period overnight. Such a temperature assisted mechanism of replication bears similarity to Polymerase Chain Reaction (PCR), a routine laboratory procedure to multiply DNA segments. Michaelian suggests that traditional origin of life research, expecting to describe the emergence of life from near-equilibrium conditions, is erroneous and that non-equilibrium conditions must be considered, in particular, the importance of entropy production to the emergence of life.

Since denaturation would be most probable in the late afternoon when the Archean sea surface temperature would be highest, and since late afternoon submarine sunlight is somewhat circularly polarized, the homochirality of the organic molecules of life can also be explained within the proposed thermodynamic framework.

Models to Explain Homochirality

Some process in chemical evolution must account for the origin of homochirality, i.e. all building blocks in living organisms having the same "handedness" (amino acids being left-handed, nucleic acid sugars (ribose and deoxyribose) being right-handed, and chiral phosphoglycerides). Chiral molecules can be synthesized, but in the absence of a chiral source or a chiral catalyst, they are formed in a 50/50 mixture of both enantiomers.

This is called a racemic mixture. Clark has suggested that homochirality may have started in space, as the studies of the amino acids on the Murchison meteorite showed L-alanine to be more than twice as frequent as its D form, and L-glutamic acid was more than 3 times prevalent than its D counterpart. It is suggested that polarised light has the power to destroy one enantiomer within the proto-planetary disk. Noyes showed that beta decay caused the breakdown of D-leucine, in a racemic mixture, and that the presence of ^{14}C, present in larger amounts in organic chemicals in the early Earth environment, could have been the cause.

Robert M. Hazen reports upon experiments conducted in which various chiral crystal surfaces act as sites for possible concentration and assembly of chiral monomer units into macromolecules. Once established, chirality would be selected for. Work with organic compounds found on meteorites tends to suggest that chirality is a characteristic of abiogenic synthesis, as amino acids show a left-handed bias, whereas sugars show a predominantly right-handed bias.

Self-organization and Replication

While features of self-organization and self-replication are often considered the hallmark of living systems, there are many instances of abiotic molecules exhibiting such characteristics under proper conditions.

For example Martin and Russel show that physical compartmentation by cell membranes from the environment and self-organization of self-contained redox reactions are the most conserved attributes of living things, and they argue therefore that inorganic matter with such attributes would be life's most likely last common ancestor.

Virus self-assembly within host cells has implications for the study of the origin of life, as it lends further credence to the hypothesis that life could have started as self-assembling organic molecules.

From Organic Molecules to Protocells

The question "How do simple organic molecules form a protocell?" is largely unanswered but there are many hypotheses. Some of these postulate the early appearance of nucleic acids ("genes-first") whereas others postulate the evolution of biochemical reactions and pathways first ("metabolism-first"). Recently, trends are emerging to create hybrid models that combine aspects of both.

"Genes First" Models: The RNA World

The RNA world hypothesis describes an early Earth with self-replicating and catalytic RNA but no DNA or proteins. This has spurred scientists to try to determine if relatively short RNA molecules could have spontaneously formed that were capable of catalysing their own continuing replication. A number of hypotheses of modes of formation have been put forward. Early cell membranes could have formed spontaneously from proteinoids, protein-like molecules that are produced when amino acid solutions are heated—when present at the correct concentration in aqueous solution, these form microspheres which are observed to behave similarly to membrane-enclosed compartments.

Other possibilities include systems of chemical reactions taking place within clay substrates or on the surface of pyrite rocks. Factors supportive of an important role for RNA in early life include its ability to act both to store information and catalyse chemical reactions (as a ribozyme); its many important roles as an intermediate in the expression and maintenance of the genetic information (in the form of DNA) in modern organisms; and the ease of chemical synthesis of at least the components of the molecule under conditions approximating the early Earth. Relatively short RNA molecules which can duplicate others have been artificially produced in the lab. Such replicase RNA, which functions as both code and catalyst provides a template upon which copying can occur. Jack Szostak has shown that certain catalytic RNAs can, indeed, join smaller RNA sequences together, creating the potential, in the right conditions for self-replication. If these were present, Darwinian selection would favour the proliferation of such self-catalysing structures, to which further functionalities could be added. Lincoln and Joyce identified an RNA enzyme capable of self sustained replication.

Researchers have pointed out difficulties for the abiotic synthesis of nucleotides from cytosine and uracil. Cytosine has a half-life of 19 days at 100 °C (212 °F) and 17,000 years in freezing water. Larralde et al., say that "the generally accepted prebiotic synthesis of ribose, the formose reaction, yields numerous sugars without any selectivity." and they conclude that their "results suggest that the backbone of the first genetic material could not have contained ribose or other sugars because of their instability." The ester linkage of ribose and phosphoric acid in RNA is known to be prone to hydrolysis.

A slightly different version of the RNA-world hypothesis is that a different type of nucleic acid, such as PNA, TNA or GNA, was the first one to emerge as a self-reproducing molecule, to be replaced by RNA only later. Pyrimidine ribonucleosides and their respective nucleotides have been prebiotically synthesised by a sequence of reactions which by-pass the free sugars, and are assembled in a stepwise fashion by going against the dogma that nitrogenous and oxygenous chemistries should be avoided.

In a series of publications, The Sutherland Group at the School of Chemistry, University of Manchester have demonstrated high yielding routes to cytidine and uridine ribonucleotides built from small 2 and 3 carbon fragments such as glycolaldehyde, glyceraldehyde or glyceraldehyde-3-phosphate, cyanamide and cyanoacetylene. One of the steps in this sequence allows the isolation of enantiopure ribose aminooxazoline if the enantiomeric excess of glyceraldehyde is 60 % or greater.

This can be viewed as a prebiotic purification step, where the said compound spontaneously crystallised out from a mixture of the other pentose aminooxazolines. Ribose aminooxazoline can then react with cyanoacetylene in a mild and highly efficient manner to give the alpha cytidine ribonucleotide. Photoanomerization with UV light allows for inversion about the 1' anomeric centre to give the correct beta stereochemistry.

In 2009 they showed that the same simple building blocks allow access, via phosphate controlled nucleobase elaboration, to 2',3'-cyclic pyrimidine nucleotides directly, which are known to be able to polymerise into RNA. This paper also highlights the possibility for the photo-sanitization of the pyrimidine-2',3'-cyclic phosphates. James Ferris's studies have shown that clay minerals of montmorillonite will catalyse the formation of RNA in aqueous solution, by joining activated mono RNA nucleotides to join together to form longer chains. Although these chains have random sequences, the possibility that one sequence began to non-randomly increase its frequency by increasing the speed of its catalysis is possible to "kick start" biochemical evolution.

"Metabolism First" Models

Several models reject the idea of the self-replication of a "naked-gene" and postulate the emergence of a primitive metabolism which could provide an environment for the later emergence of RNA replication.

Iron-sulfur World

One of the earliest incarnations of this idea was put forward in 1924 with Alexander Oparin's notion of primitive self-replicating vesicles which predated the discovery of the structure of DNA. More recent variants in the 1980s and 1990s include Günter Wächtershäuser's iron-sulfur world theory and models introduced by Christian de Duve based on the chemistry of thioesters.

However, the idea that a closed metabolic cycle, such as the reductive citric acid cycle, could form spontaneously (proposed by Günter Wächtershäuser) remains debated. In an article entitled "Self-Organizing Biochemical Cycles", the late Leslie Orgel summarized his analysis of the proposal by stating, "There is at present no reason to expect that multistep cycles such as the reductive citric acid cycle will self-organize on the surface of FeS/FeS2 or some other mineral." It is possible that another type of metabolic pathway was used at the beginning of life.

For example, instead of the reductive citric acid cycle, the "open" acetyl-CoA pathway (another one of the five recognised ways of carbon dioxide fixation in nature today) would be compatible with the idea of self-organisation on a metal sulfide surface. The key enzyme of this pathway, carbon monoxide dehydrogenase/acetyl-CoA synthase harbours mixed nickel-iron-sulfur clusters in its reaction centres and catalyses the formation of acetyl-CoA (which may be regarded as a modern form of acetyl-thiol) in a single step.

Thermosynthesis World

Today's bioenergetic process of fermentation is related to the just mentioned citric acid cycle or the Acetyl-CoA pathway that have been connected to the primordial iron-sulfur world. In a different approach, today's bioenergetic process of chemiosmosis, which plays an essential role in cellular respiration and photosynthesis, is considered as more fundamental than fermentation: in Anthonie Muller's "thermosynthesis world" the ATP Synthase enzyme that sustains chemiosmosis is proposed as today's enzyme that is the closest connected to the first metabolic process.

First life needed an energy source to bring about the condensation reaction that yielded the peptide bonds of proteins and the phosphodiester bonds of RNA. In a generalization and thermal variation of the binding change mechanism of today's ATP Synthase, the "First Protein" would have bound substrates (peptides, phosphate, nucleosides, RNA 'monomers') and condensed them to a reaction product that remained bound until it after a temperature change was released upon a thermal unfolding.

The energy source of the thermosynthesis world was thermal cycling, the result of suspension of the protocell in a convection current, as is plausible in a volcanic hot spring; the convection accounts for the self-organization and dissipative structure required in any origin of life model. The still ubiquitous role of thermal cycling in germination and cell division is considered a relic of primordial thermosynthesis.

By phosphorylating cell membrane lipids, this 'First Protein' gave a selective advantage to the lipid protocell that contained the protein. In the beginning this First Protein also synthesized a library with many proteins, of which only a minute fraction had thermosynthesis capabilities. Just as proposed by Dyson for the first proteins, the First Protein propagated functionally: it made daughters with similar capabilities, but it did not copy itself. Functioning daughters consisted of different amino acid sequences.

Over a long time, RNA sequences were selected among the at first randomly synthesized RNAs by the criterion of speed and efficiency increase of First Protein synthesis, for instance by the creation of RNA that functioned as messenger RNA, Transfer RNA and ribosomal RNA, or, even more generally, all the components of the RNA World were also generated and selected. The thermosynthesis world therefore in theory accounts for the origin of the genetic machinery.

Whereas the iron-sulfur world identifies a circular pathway as the most simple—and therefore assumes the existence of enzymes—the thermosynthesis world does not even invoke a pathway, and does not assume the existence of regular enzymes: ATP Synthase's binding change mechanism resembles a physical adsorption process that yields free energy, rather than a regular enzyme's mechanism, which decreases the free energy.

The RNA World also implies the existence of several enzymes. But even the emergence of a single enzyme by chance is implausible. The thermosynthesis world is therefore more simple, and thus more plausible, than the iron-sulfur and RNA worlds.

Possible Role of Bubbles

Waves breaking on the shore create a delicate foam composed of bubbles. Winds sweeping across the ocean have a tendency to drive things to shore, much like driftwood collecting on the beach. It is possible that organic molecules were concentrated on the shorelines in much the same way. Shallow coastal waters also tend to be warmer, further concentrating the molecules through evaporation. While bubbles composed mostly of water burst quickly, water containing amphiphiles forms much more stable bubbles, lending more time to the particular bubble to perform these crucial reactions.

Amphiphiles are oily compounds containing a hydrophilic head on one or both ends of a hydrophobic molecule. Some amphiphiles have the tendency to spontaneously form membranes in water. A spherically closed membrane contains water and is a hypothetical precursor to the modern cell membrane. If a protein would increase the integrity of its parent bubble, that bubble had an advantage, and was placed at the top of the natural selection waiting list. Primitive reproduction can be envisioned when the bubbles burst, releasing the results of the 'experiment' into the surrounding medium. Once enough of the 'right stuff' was released into the medium, the development of the first prokaryotes, eukaryotes, and multicellular organisms could be achieved.

Similarly, bubbles formed entirely out of protein-like molecules, called microspheres, will form spontaneously under the right conditions. But they are not a likely precursor to the modern cell membrane, as cell membranes are composed primarily of lipid compounds rather than amino-acid compounds.

A recent model by Fernando and Rowe suggests that the enclosure of an autocatalytic non-enzymatic metabolism within protocells may have been one way of avoiding the side-reaction problem that is typical of metabolism first models.

Other Models

Autocatalysis

In 1993 Stuart Kauffman proposed that life initially arose as autocatalytic chemical networks.

British ethologist Richard Dawkins wrote about autocatalysis as a potential explanation for the origin of life in his 2004 book *The Ancestor's Tale*. Autocatalysts are substances which catalyse the production of themselves, and therefore have the property of being a simple molecular replicator. In his book, Dawkins cites experiments performed by Julius Rebek and his colleagues at the Scripps Research Institute in California in which they combined amino adenosine and pentafluorophenyl ester with the autocatalyst amino adenosine triacid ester (AATE). One system from the experiment contained variants of AATE which catalysed the synthesis of themselves. This experiment demonstrated the possibility that autocatalysts could exhibit competition within a population of entities with heredity, which could be interpreted as a rudimentary form of natural selection.

Clay Theory

A model for the origin of life based on clay was forwarded by A. Graham Cairns-Smith of the University of Glasgow in 1985 and explored as a plausible illustration by several other scientists, including Richard Dawkins. Clay theory postulates that complex organic molecules arose gradually on a pre-existing, non-organic replication platform—silicate crystals in solution. Complexity in companion molecules developed as a function of selection pressures on types of clay crystal is then exapted to serve the replication of organic molecules independently of their silicate "launch stage".

Cairns-Smith is a staunch critic of other models of chemical evolution. However, he admits that like many models of the origin of life, his own also has its shortcomings (Horgan 1991).

In 2007, Kahr and colleagues reported their experiments to examine the idea that crystals can act as a source of transferable information, using crystals of potassium hydrogen phthalate. "Mother" crystals with imperfections were cleaved and used as seeds to grow "daughter" crystals from solution. They then examined the distribution of imperfections in the crystal system and found that the imperfections in the mother crystals were indeed reproduced in the daughters. The daughter crystals had many additional imperfections. For a gene-like behaviour the additional imperfections should be much less than the parent ones, thus Kahr concludes that the crystals "were not faithful enough to store and transfer information from one generation to the next".

Gold's "Deep-hot Biosphere" Model

In the 1970s, Thomas Gold proposed the theory that life first developed not on the surface of the Earth, but several kilometers below the surface. The discovery in the late 1990s of nanobes (filamental structures that are smaller than bacteria, but that may contain DNA) in deep rocks might be seen as lending support to Gold's theory.

It is now reasonably well established that microbial life is plentiful at shallow depths in the Earth, up to 5 kilometres (3.1 mi) below the surface, in the form of extremophile archaea, rather than the better-known eubacteria (which live in more accessible conditions). It is claimed that discovery of microbial life below the surface of another body in our solar system would lend significant credence to this theory. Thomas Gold also asserted that a trickle of food from a deep, unreachable, source is needed for survival because life arising in a puddle of organic material is likely to consume all of its food and become extinct. Gold's theory is that flow of food is due to out-gassing of primordial methane from the Earth's mantle; more conventional explanations of the food supply of deep microbes (away from sedimentary carbon compounds) is that the organisms subsist on hydrogen released by an interaction between water and (reduced) iron compounds in rocks.

"Primitive" Extraterrestrial Life

An alternative to Earthly abiogenesis is the hypothesis that primitive life may have originally formed extraterrestrially, either in space or on a nearby planet (Mars). (Note that exogenesis is related to, but not the same as, the notion of panspermia). A supporter of this theory was Francis Crick.

Organic compounds are relatively common in space, especially in the outer solar system where volatiles are not evaporated by solar heating. Comets are encrusted by outer layers of dark material, thought to be a tar-like substance composed of complex organic material formed from simple carbon compounds after reactions initiated mostly by irradiation by ultraviolet light. It is supposed that a rain of material from comets could have brought significant quantities of such complex organic molecules to Earth.

An alternative but related hypothesis, proposed to explain the presence of life on Earth so soon after the planet had cooled down, with apparently very little time for prebiotic evolution, is that life formed first on early Mars. Due to its smaller size Mars cooled before Earth (a difference of hundreds of millions of years), allowing prebiotic processes there while Earth was still too hot. Life was then transported to the cooled Earth when crustal material was blasted off Mars by asteroid and comet impacts. Mars continued to cool faster and eventually became hostile to the continued evolution or even existence of life (it lost its atmosphere due to low volcanism); Earth is following the same fate as Mars, but at a slower rate.

Neither hypothesis actually answers the question of how life first originated, but merely shifts it to another planet or a comet. However, the advantage of an extraterrestrial origin of primitive life is that life is not required to have evolved on each planet it occurs on, but rather in a single location, and then spread about the galaxy to other star systems via cometary and/or meteorite impact. Evidence to support the hypothesis is scant, but it finds support in recent study of Martian meteorites found in Antarctica and in studies of extremophile microbes. Additional support comes from a recent discovery of a bacterial ecosystem whose energy source is radioactivity.

A 2001 experiment led by Jason Dworkin subjected a frozen mixture of water, methanol, ammonia and carbon monoxide to UV radiation, mimicking conditions found in an extraterrestrial environment. This combination yielded large amounts of organic material that self-organised to form bubbles or micelles when immersed in water. Dworkin considered these bubbles to resemble cell membranes that enclose and concentrate the chemistry of life, separating their interior from the outside world.

The bubbles produced in these experiments were between 10 to 40 micrometres (0.00039 to 0.0016 in), or about the size of red blood cells. Remarkably, the bubbles fluoresced, or glowed, when exposed to UV light. Absorbing UV and converting it into visible light in this way was considered one possible way of providing energy to a primitive cell. If such bubbles played a role in the origin of life, the fluorescence could have been a precursor to primitive photosynthesis. Such fluorescence also provides the benefit of acting as a sunscreen, diffusing any damage that otherwise would be inflicted by UV radiation. Such a protective function would have been vital for life on the early Earth, since the ozone layer, which blocks out the sun's most destructive UV rays, did not form until after photosynthetic life began to produce oxygen.

Extraterrestrial Amino Acids

Another idea is that amino acids which were formed extraterrestrially arrived on Earth via comets. In 2009 it was announced by NASA that scientists have identified one of the fundamental chemical buildings blocks of life in a comet for the first time: glycine, an amino acid, was detected in the material ejected from Comet Wild-2 in 2004 and grabbed by NASA's Stardust probe.

Tiny grains, just a few thousandths of a millimetre in size, were collected from the comet and returned to Earth in 2006 in a sealed capsule, and distributed among the world's leading astro-biology labs. NASA said in a statement that it took some time for the investigating team, led by Dr Jamie Elsila, to convince itself that the glycine signature found in Stardust's sample bay was genuine and not just Earthly contamination.

Glycine has been detected in meteorites before and there are also observations in interstellar gas clouds claimed for telescopes, but the Stardust find is described as a first in cometary material. It is known that prior to the emergence of life on Earth, the early solar system's planets were regularly bombarded by comets. Dr. Carl Pilcher, who leads NASA's Astrobiology Institute commented that "The discovery of glycine in a comet supports the idea that the fundamental building blocks of life are prevalent in space, and strengthens the argument that life in the Universe may be common rather than rare."

Lipid World

This theory postulates that the first self-replicating object was lipid-like. It is known that phospholipids form bilayers in water while under agitation– the same structure as in cell membranes. These molecules were not present on early Earth, however other amphiphilic long chain molecules also form membranes.

Furthermore, these bodies may expand (by insertion of additional lipids), and under excessive expansion may undergo spontaneous splitting which preserves the same size and composition of lipids in the two progenies. The main idea in this theory is that the molecular composition of the lipid bodies is the preliminary way for information storage, and evolution led to the appearance of polymer entities such as RNA or DNA that may store information favorably. Still, no biochemical mechanism has been offered to support the Lipid World theory.

Polyphosphates

The problem with most scenarios of abiogenesis is that the thermodynamic equilibrium of amino acid versus peptides is in the direction of separate amino acids. What has been missing is some force that drives polymerization. The resolution of this problem may well be in the properties of polyphosphates. Polyphosphates are formed by polymerization of ordinary monophosphate ions PO_4^{-3}. Several mechanisms for such polymerization have been suggested. Polyphosphates cause polymerization of amino acids into peptides. They are also logical precursors in the synthesis of such key biochemical compounds as ATP. A key issue seems to be that calcium reacts with soluble phosphate to form insoluble calcium phosphate (apatite), so some plausible mechanism must be found to keep calcium ions from causing precipitation of phosphate. There has been much work on this topic over the years, but an interesting new idea is that meteorites may have introduced reactive phosphorus species on the early Earth.

PAH World Hypothesis

Other sources of complex molecules have been postulated, including extraterrestrial stellar or interstellar origin. For example, from spectral analyses, organic molecules are known to be present in comets and meteorites. In 2004, a team detected traces of polycyclic aromatic hydrocarbons (PAH's) in a nebula. Those are the most complex molecules so far found in space. The use of PAH's has also been proposed as a precursor to the RNA world in the PAH world hypothesis. The Spitzer Space Telescope has recently detected a star, HH 46-IR, which is forming by a process similar to that by which the sun formed. In the disk of material surrounding the star, there is a very large range of molecules, including cyanide compounds, hydrocarbons, and carbon monoxide. PAHs have also been found all over the surface of galaxy M81, which is 12 million light years away from the Earth, confirming their widespread distribution in space.

Multiple Genesis

Different forms of life may have appeared quasi-simultaneously in the early history of Earth. The other forms may be extinct, leaving distinctive fossils through their different biochemistry (e.g., using arsenic instead of phosphorus), survive as extremophiles, or simply be unnoticed through their being analogous to organisms of the current life tree. Hartman for example combines a number of theories together, by proposing that:

The first organisms were self-replicating iron-rich clays which fixed carbon dioxide into oxalic and other dicarboxylic acids. This system of replicating clays and their metabolic phenotype then evolved into the sulfide rich region of the hotspring acquiring the ability to fix nitrogen. Finally phosphate was incorporated into the evolving system which allowed the synthesis of nucleotides and phospholipids. If biosynthesis recapitulates biopoesis, then the synthesis of amino acids preceded the synthesis of the purine and pyrimidine bases. Furthermore the polymerization of the amino acid thioesters into polypeptides preceded the directed polymerization of amino acid esters by polynucleotides. Lynn Margulis's endosymbiotic theory suggests that multiple forms of bacteria entered into symbiotic relationship to form the eukaryotic cell. The horizontal transfer of genetic material between bacteria promotes such symbiotic relationships, and thus many separate organisms may have contributed to building what has been recognised as the Last Universal Common Ancestor (LUCA) of modern organisms. James Lovelock's Gaia theory, proposes that such bacterial symbiosis establishes the environment as a system produced by and supportive of life. His arguments strongly weaken the case for life having evolved elsewhere in the solar system.

History

The first organic cofactor to be discovered was NAD^+, which was identified by Arthur Harden and William Youndin 1906. They noticed that adding boiled and filtered yeast extract greatly accelerated alcoholic fermentation in unboiled yeast extracts. They called the unidentified factor responsible for this effect a *coferment*. Through a long and difficult purification from yeast extracts, this heat-stable factor was identified as a nucleotide sugar phosphate by Hans von Euler-Chelpin. Other cofactors were identified throughout the early 20th century, with ATP being isolated in 1929 by Karl Lohmann, and coenzyme A being discovered in 1945 by Fritz Albert Lipmann. The functions of these molecules were at first mysterious, but in 1936, Otto Heinrich Warburg identified the function of NAD^+ in hydride transfer. This discovery was followed in the early 1940s by the work of Herman Kalckar, who established the link between the oxidation of sugars and the generation of ATP. This confirmed the central role of ATP in energy transfer that had been proposed by Fritz Albert Lipmann in 1941. Later, in 1949, Morris Friedkin and Albert L. Lehninger proved that NAD^+ linked metabolic pathways such as the citric acid cycle and the synthesis of ATP.

Non-enzymatic Cofactors

The term is used in other areas of biology to refer more broadly to non-protein (or even protein) molecules that either activate, inhibit or are required for the protein to function. For example, ligands such as hormones that bind to and activate receptor proteins are termed cofactors or coactivators, while molecules that inhibit receptor proteins are termed corepressors.

Coenzymes of Oxidation Reduction Reactions

Coenzymes are small organic molecules that link to enzymes and whose presence is essential to the activity of those enzymes. Coenzymes belong to the larger group called cofactors, which also includes metal ions; cofactor is the more general term for small molecules required for the activity of their associated enzymes. The relationship between these two terms is as follows

I. Cofactors:
- Essential ions
- Loosely bound (forming metal-activated enzymes)
- Tightly bound (forming metalloenzymes
- Coenzymes
- Tightly bound prosthetic groups
- 2 Loosely bound cosubstrates.

Many coenzymes are derived from vitamins. Table 1 lists vitamins, the coenzymes derived from them, the type of reactions in which they participate, and the class of coenzyme.

Prosthetic groups are tightly bound to enzymes and participate in the catalytic cycles of enzymes. Like any catalyst, an enzyme–prosthetic group complex undergoes changes during the reaction, but before it can catalyse another reaction, it must return to its original state.

Flavin adenine dinucleotide (FAD) is a prosthetic group that participates in several intracellular oxidation-reduction reactions. During the catalytic cycle of the enzyme succinate dehydrogenase, FAD accepts two electrons from succinate, yielding fumarate as a product. Because FAD is tightly bound to the enzyme, the reaction is sometimes shown this way

succinate + E–FAD → fumarate + E–FADH$_2$

where E–FAD stands for the enzyme tightly bound to the FAD prosthetic group. In this reaction the coenzyme FAD is reduced to FADH$_2$ and remains tightly bound to the enzyme throughout. Before the enzyme can catalyse the oxidation of another succinate molecule, the two electrons now belonging to E–FADH$_2$ must be transferred to another electron acceptor, ubiquinone. The regenerated E–FAD complex can then oxidize another succinate molecule.

Cosubstrates are loosely bound coenzymes that are required in stoichiometric amounts by enzymes. The molecule nicotinamide adenine dinucleotide (NAD) acts as a cosubstrate in the oxidation-reduction reaction that is catalysed by malate dehydrogenase, one of the enzymes of the citric acid cycle.

malate + NAD$^+$ → oxaloacetate + NADH

In this reaction, malate and NAD$^+$ diffuse into the active site of malate dehydrogenase. Here NAD$^+$ accepts two electrons from malate; oxaloacetate and NADH then diffuse out of the active site. The reduced NADH must then be returned to its NAD$^+$ form. For each catalytic cycle, a "new" NAD$^+$ molecule is needed if the reaction is to occur; thus, stoichiometric quantities of the cosubstrate are needed. The reduced form of this coenzyme (NADH) is converted back to the oxidized form (NAD$^+$) via a number of simultaneously occurring processes in the cell, and the regenerated NAD$^+$ can then participate in another round of catalysis.

Coenzymes, then, are a type of cofactor. They are small organic molecules that bind tightly (prosthetic groups) or loosely (cosubstrates) to enzymes as they participate in catalysis.

Population Nutrition Health Promotion and Government Policy

What is Public Health?

Public health is a linked system of federal, state and local (city and county) health departments, operated by the government and given the authority and responsibility to monitor and protect the public's health.

The mission of public health is "to assure conditions in which people can be healthy" (Institute of Medicine, The Future of Public Health).

Since nutrition is an essential aspect of the conditions in which people can be healthy, public health nutrition is part of the public health system.

What is Public Health Nutrition?

- Strives to improve or maintain optimum nutritional health of the whole population and high risk or vulnerable subgroups within the population.
- Emphasizes health promotion and disease prevention but may include therapeutic and rehabilitative services when these needs are not adequately addressed by other parts of the health care system.
- Uses multiple, coordinated strategies to reach and influence the community, and organizations and individuals that make up the community.
- Requires organized and integrated community nutrition efforts with leadership provided by the state and local health agency.

Community nutrition efforts involve a wide range of programs that provide increased access to food resources, nutrition information and education, and health-related care. They also include efforts to change behaviour and environments and to initiate policy.

What types of organizations do this kind of work?
- Many types of organizations are involved in public health/community nutrition work.
- Leadership of community nutrition efforts is usually provided by a public health nutritionist employed in an "official" public health agency—a state, city, or county health department.
- Public-private partnerships or coalitions are frequently formed to address priority nutrition problems in the community.
- Ideally, organizations providing nutrition-related programs communicate and coordinate to effectively address nutrition problems and avoid service gaps.

What's the difference between community nutrition and public health nutrition?

The term community nutrition is often used to reflect the wide range of delivery settings and sponsoring organizations for nutrition-related programs and services. Community nutrition services tend to be directed to individuals and groups in the community. The term public health nutrition has historically been used for the responsibilities carried out by health departments at local, state and federal levels. The programs offered by public health agencies are usually directed to communities, organizations and systems and have as their goal health promotion and disease prevention. Additionally, public health nutrition is often involved in policy development. In practice the terms public health nutrition and community nutrition tend to be used interchangeably.

Why is it important to know about public health nutrition?

Adequate nutrition for all is the goal: Adequate food and balanced nutrient intake are basic necessities for life, health and well being. Nutrition affects health from conception to old age. Adequate nutrition is especially important in periods of rapid growth and development. Poor nutrition during pregnancy, infancy, childhood and adolescence can mean stunted physical, mental and social development with lifelong consequences. Chronic dietary deficiency, excess or imbalance predisposes individuals to or aggravates a spectrum of disease conditions, and ultimately affect the quality and length of life.

Dietary factors are associated with five of the ten leading causes of death: Coronary heart disease, some types of cancer, stroke, non-insulin dependent diabetes (type 2 diabetes), and atherosclerosis are associated with dietary factors. Dietary excesses and imbalances contribute to the development of these diseases. Currently attention is focused on total caloric intake; amount and type of fat; vitamins such as folic acid and the antioxidants of vitamins A, C and E; minerals such as calcium; and other nutritive substances such as fiber and flavonoids. Overweight and obesity which are estimated to affect over a third of the population is also an important contributing factor for disease and disability.

Maternal and child nutrition sets the stage for life: The health of mothers and infants has historically been a focus of public health and public health nutrition. Balanced diet and appropriate weight gain have received attention in the past. Now attention is also directed to preconceptual concerns such as folic acid intake and its association with neural tube defects. Recent research links factors in the fetal environment to risk for adult diseases including diabetes and cancer. Breastfeeding for the first year of life is recommended because of its many benefits to infants and their mothers. Childhood is a time when food preferences and habits are shaped. Childhood nutrition affects growth and development, immune status, and social and cognitive ability. The nutritional intake of children with special health care needs also requires close scrutiny. Low calcium intake of girls and young women sets the stage for osteoporosis in later years.

Vulnerable subgroups are at high risk for nutritional problems: Some subgroups of the population, including people with low incomes, some racial and ethnic minority groups, and people with disabilities (defined as functional impairments) experience a disproportionate amount of preventable illness and premature death. Nutrition is an important contributing factor. Some groups, especially those who are economically disadvantaged or isolated, experience periodic or chronic hunger (also called food insecurity) resulting in undernutrition.

Reaching these groups with accessible, culturally-relevant, nutrition programs and services presents aspecial challenge to public health agencies and all community nutrition providers.

Targeting vulnerable subgroups and designing programs to meet their special needs is a strategy used by public health to attempt to reduce disparities in nutritional status and health among population subgroups.

Behaviour change is challenging: Nutrition behaviour (including food selection, preparation and consumption) is the product of culture, education, economics, food availability, social strata, family position and health status. Nutritional status depends on all those factors plus biological and genetic factors. Guiding all members of the population toward more healthful food choices and optimum nutritional health is a great challenge. And doing so early enough to prevent the development of disease is a goal of public health nutrition. Meeting this challenge requires the use of multiple, reinforcing behaviour change strategies, including food and nutrition information and education. Other strategies include:

— structuring the environment to enable positive food choices (e.g., juice machines replace pop machines)
— modifying food ingredients and preparation techniques to reduce fat content
— improving the availability of foods such as fruits and vegetables, and
— enacting legislation and regulation (such as required nutrition labels on food packages).

These indirect, environmental strategies complement the direct, individualized nutrition education and counseling provided in health care and other settings.

What Organizations Are Involved in Community Nutrition Work?

A wide ranging network of providers (including organizations, professionals and volunteers) help assure that needed community nutrition programs and services are available to the population and targeted subgroups. The combined efforts of many providers are necessary to meet the challenge of assuring optimum nutrition for all. A range of providers also enables nutrition programs to be more closely tailored to the needs and interests of the target population or subgroup so programs are easily accessible and culturally-relevant to their users. A part of the coordinating role of the public health departments is to assure necessary and appropriate nutrition services are available and duplication is avoided.

Where Do You Find Community Nutrition Professionals?

Government agencies:
- public health department (WIC, maternal and child health, health promotion and disease prevention units)
- cooperative extension

- social/human service agencies

Nutrition and food assistance programs:
- Food Stamp Program
- Elderly Nutrition Program (congregate meals and home delivered meals)
- food shelves and emergency meals programs

Schools:
- school breakfast and lunch
- student health services
- nutrition and health education curriculum

Day care centres and Head Start programs:
- Child and Adult Care Food Program.

Community not-for-profit Organizations

Voluntary Health Organizations:
- March of Dimes Foundation
- American Heart Association
- American Cancer Society.

Business and Industry:
- Food corporations
- Food producers and commodity groups.

Health care organizations:
- community health centres
- hospitals
- clinics
- managed care organizations (HMOs)
- home health agencies.

Who are the community nutrition professionals in your community? How to locate the community nutrition professionals in your community?

- Look for the organizations listed above in the government and business sections of the phone book
- Call the local health department
- Call the state Dietetic Association
- Search for the organization's Web page on the internet.

Bibliography

Abdullah, A Mohamed : *Food Security and Gender Inequality*, Abjiheet, Delhi, 2008.

Alec, William: *The Dairy Chemical Industry*, London: Longman Group Limited, 1971.

Arora, Dinesh: *Biotech's Dictionary of Dairy Science*, Biotech Books, Delhi, 296.

Avanish K. Tiwari: *Food Security and Global Economy*, Pentagon Press, Delhi, 2009.

Babita Bohra: *Dairy Farming in Mountain Areas*, Daya, Delhi, 2006.

Basavaraj S. Benni, Rawat: *Dairy Co-operative Management and Practice*, Delhi, 2005.

Bhattacharya, Lata: *Biochemistry of Nutrition*, Discovery, Delhi, 2010.

Bhutani, R.C.: *Fruit and Vegetable Preservation*, Biotech Books, Delhi, 2003.

Bohra, Babita: *Dairy Farming in Mountain Areas*, Daya, Delhi, 2006.

Brij K. Taimni: *Food Security in 21 Century : Perspective and Vision*, Konark, Delhi, 2001.

Brock, H. : *History of Dairy Chemistry*, New York: Norton, 1992.

Bucciarelli, L.: *Designing Engineers in Dairy*, Cambridge: MIT Press, 1995.

Bushnell, R.B. : *Dry Cow Feeding and Management*, A Western Regional Extension Publication, 1979.

Chakraborty, Sudip : *Food Security and Child Labour : The Case of a Hazardous Occupation*, Deep and Deep, Delhi, 2011.

Chand, Ram: *Decision-Making of Dairy Beneficiaries: Role of Aspiration, Motivation and Knowledge*, Om Pub, Delhi, 2010.

Chaturvedi, Pradeep : *Food Security and Panchayati Raj*, Concept, Delhi, 1997.

Chhazllani, V K : *Dairy Chemistry and Animal Nutrition*, Manglam Pub, Delhi, 2008.

Cohen, Lizabeth, *Making a New Deal, Industrial Workers in Chicago, 1919-1939* Cambridge University Press, 1991.

Collymore L.: *Fruit Production in Barbados*, Port of Spain, Trinidad and Tobago, 1996.

Cristobal Noe Aguilar : *Food Science and Food Biotechnology in Developing Countries,* Asiatech Pub, 2008.

Damasio, AR.: *Descarte's Error: Emotion, Reason and the Human Brain,* New York, 1994.

David, M.: *Ideas in Chemistry: A History of the Science,* New Brunswick, N.J.: Rutgers University Press, 1992.

De, Sukumar: *Outlines of Dairy Technology,* Oxford University Press, Delhi, 2001.

Devraj B: *Impact of Dairy On Small and Marginal Farmers,* Prateeksha Publications, Delhi, 2010.

Dinesh Arora: *Biotech's Dictionary of Dairy Science,* Biotech Books, Delhi, 296.

Droop, H. Richmond: *Laboratory Manual of Dairy Analysis,* Biotech, 2004.

Fox, Patrick F.: *Advanced Dairy Chemistry: Proteins,* New York: Elsevier Applied Science, 1992.

Friberg, Stig. E..: *Food Emulsions,* New York: M. Dekker, 1997.

Guarti, Luigi, *The Valuation of Firms,* Blackwell Publishing, 1994.

Gunjan Goel : *Applied Dairy and Food Microbiology,* Agrotech, 2005.

Hui, Yiu H. : *Dairy Science and Technology Handbook,* New York: Wiley, 1993.

Jacobson, M. : *Safe Food: Eating Wisely in a Risky World,* Washington, DC: Living Planet Press, 1991.

Jensen, Robert G. : *Handbook of Milk Composition,* San Diego: Academic Press, 1995.

Jha, S N : *Dairy and Food Processing Plant Maintenance : Theory and Practice,* International Book Distributing, Delhi, 2006.

John Prince: *Dairy Farming: Being the Theory, Practice, and Methods of Dairying,* New York, 1888.

Johnston, J. R. : *Molecular Genetics of Yeast, a Practical Approach.* IRL Press, Oxford 1994.

Kango, Mangala: *Normal Nutrition: Fundamental and Management,* RBSA, Delhi, 2003.

Kapoor, Ajay : *Dairy Science and Technology,* Vishvabharti Pub, Delhi, 2005.

Koli, P. A.: *Dairy Development in India: Challenges Before Co-Operatives,* Shruti Pub, Delhi, 2007.

Krimsky, Sheldon : *Biotechnics and Society: The Rise of Industrial Genetics,* New York: Praeger Publishers, 1991.

Kumar, Shashi : *Biodiversity and Food Security,* Atlantic, Delhi, 2002.

Law, Barry A.: *Microbiology and Biochemistry of Cheese and Fermented Milk,* London: Blackie Academic & Professional, 1997.

Leena Parihar: *Dairy Microbiology*, Agrobios, Delhi, 2008.

Michael, J. Lewis,: *SeparationProcesses in the Food and Biotechnology Industries*, Cambridge, U.K.: Woodhead, 1996.

Modi, H.A. : *Dairy Microbiology*, Aavishkar Publishers, Delhi, 2009.

Mohana Swamy : *Agro's Dictionary of Dairy Science*, Agro Botanical, 1995.

Mowery, D. C. : *Technology and Wealth of Nations*, Stanford: Stanford University Press, 1992.

Mudgal, V. D.; K. K. Singhal and D. D. Sharma: *Advances in Dairy Animal Production*, International Book Distribut, 2003.

Narang, R.K.: *Fruit and Vegetable Preservation Techniques*, APH Pub, Delhi, 2010.

Pandey, D. N. and Amita Bajpai: *Recent Trends in Animal Nutrition and Feed Technology for Livestock, Pets and Laboratory Animals*, International, 2003.

Parihar, Pradeep and Leena Parihar: *Dairy Microbiology*, Agrobios, Delhi, 2008.

Patton, Stuart: *Principles of Dairy Chemistry*, Huntington, N.Y.: Krieger, 1976.

Paul B. : *Food Biotechnology in Ethical Perspective*, Aspen, CO: Aspen Publishers, 1997.

Pirtle, Thomas Ross: *History of the Dairy Industry*, Chicago: Mojonnier Bros. Company, 1926.

Pradeep Chaturvedi: *Food Security in South Asia*, Concept, Delhi, 2002.

Qystein V. Sjaastad: *Physiology of Domestic Animals*, International Book Distributing Co., Delhi, 2005.

Ram Chand: *Decision-Making of Dairy Beneficiaries : Role of Aspiration, Motivation and Knowledge*, Om Pub, Delhi, 2010.

Ramakant Sharma : *Chemical and Microbiological Analysis of Milk and Milk Products*, International Book Distributing, 2006.

Rao, M K : *Food and Dairy Microbiology*, Manglam Pub, Delhi, 2007.

Rao, P. Venkateshwara : *Dairy Farm Business Management*, Biotech Books, Delhi, 2008.

Saini, M.L. : *Plant Breeding and Crop Improvement*, CBS, Delhi, 1997.

Samvel, A. P. V.: *Agri-Business Management*, Satish Serial Pub, Delhi, 2008.

Sarkar, A: *Advanced Organic Chemistry: Reactions and Mechanisms*, Swastik Publications, Delhi, 2011.

Shepherd, D. : *Homeopathy for the First Aider*, Sussex, England: Health Science Press, 1953.

Shukla, Arvind N.: *Textbook of Dairy Chemistry*, Discovery Pub, Delhi 2010.

Singh, Gajendra : *Food for All : An Assessment of Food Security in Indian Context*, MD Pub, Delhi, 2007.

Singh, Harmeet: *Dairy Farming*, APH, Delhi, 2005.

Singh, S K : *Biotechnology, Plant Propagation and Plant Breeding*, Campus Books, Delhi, 2008.

Singh, Vir and Babita Bohra: *Dairy Farming in Mountain Areas*, Daya, Delhi, 2006.

Singhal, K K and D D Sharma : *Advances in Dairy Animal Production*, International Book Distribut, 2003.

Spreer, Edgar : *Milk and Dairy Product Technology*, Translated by Axel Mixa. New York: M. Dekker, 1998.

Sriram Sridhar : *Enzyme and Food Biotechnology*, Wisdom Press, 2011.

Sujata K. Dass: *Biotechnology and Food Security*, Isha Books, Delhi, 2004.

Sukumar De: *Outlines of Dairy Technology*, Oxford University Press, Delhi, 2001.

Suri, Nitin : *Molecular Biology and Biochemistry*, Oxford Book Company, Delhi, 2010.

Susanna Hornig: *A Grain of Truth: The Media, the Public, and Biotechnology*, Lantham, MD: Rowman and Littlefield, 2001.

Thompson, Paul B. : *Food Biotechnology in Ethical Perspective*, Aspen, CO: Aspen Publishers, 1997.

Thomson, Sutherland: *Grading Dairy Produce*, Medi World Press, Delhi, 1995.

Tomar, S.K. and Gunjan Goel : *Applied Dairy and Food Microbiology*, Agrotech, 2005.

Tramontano, A., : *Antibodies as enzymes*, Trends in Biochemical Sciences, 1987.

Tripathy, S.N. : *Food Biotechnology*, Dominant, 2004.

Tyagi, Prasum: *A Textbook of Animal Physiology*, Dominant, Delhi, 2010.

Upadhyay, K.G. and Vyas, S.H.: *Composition of Camel's Milk*, Gujarat Agric. University, 1982.

Venkateshwara Rao: *Dairy Farm Business Management*, Biotech Books, Delhi, 2008.

Walstra, Pieter, and Robert Jenness: *Dairy Chemistry and Physics*, New York: Wiley, 1984.

William Alec : *The Dairy Chemical Industry*, London: Longman Group Limited, 1971.

Yegge, Wilbur M., *A Basic Guide for Valuing a Company*, New York, Wiley, 1996.

Index

A

Abiogenesis, 243, 244, 246, 250, 262, 264.
Agricultural Environments, 11.
Agriculture, 1, 12, 15, 19, 37, 38, 39, 40, 42, 52, 54, 91, 92, 94, 101, 110, 124, 127.
Amino Acid, 47, 102, 121, 125, 141, 143, 144, 161, 162, 163, 234, 235, 236, 256, 259, 263, 264, 265.
Anaerobic Digestion, 148.
Animal Source Foods, 13, 14.
Antibiotics, 104.
Authority, 36, 38, 39, 40, 133, 227, 228, 268.

B

Biotechnology, 61, 83, 84, 85, 86, 87, 88, 89, 90, 91, 92, 93, 94, 95, 96, 97, 101, 102, 103, 114, 115, 116, 117, 124, 127, 128, 129, 130, 131, 134, 153, 196.
Boiling Stages, 25.
Bottling Wine, 179.
Bromelain, 137, 156, 157.
Butter, 3, 18, 21, 23, 40, 74, 122.

C

Calcium, 14, 22, 63, 75, 100, 118, 142, 154, 155, 157, 182, 183, 184, 194, 199, 200, 201, 202, 203, 204, 205, 206, 207, 208, 209, 210, 211, 212, 213, 215, 218, 224, 225, 226, 239, 240, 264, 265, 270.
Calcium Metabolism, 207, 210.
Carbohydrates, 1, 3, 4, 16, 20, 21, 57, 101, 118, 123, 124, 130, 133, 185.
Cellaring Wine, 179.
Chemical Composition, 181.
Commercial Production, 151, 165, 205.
Commercial Trade, 52.
Communications, 129.
Community Health Centres, 272.
Consumer Labelling, 44.
Consumer Protection, 38, 39.
Consumers, 3, 4, 5, 6, 7, 8, 9, 10, 12, 37, 42, 45, 53, 54, 83, 87, 90, 91, 92, 93, 94, 95, 96, 97, 98, 100, 101, 102, 105, 106, 107, 108, 113, 114, 117, 118, 125, 127, 197.
Controversy, 94.
Cooking, 15, 16, 18, 19, 20, 21, 22, 23, 24, 25, 26, 27, 28, 29, 30, 31, 32, 33, 37, 43, 44, 45, 46, 47, 48, 49, 55, 56, 119, 120, 121, 190, 194, 225, 229.
Cream, 2, 152, 170, 174.
Curd, 75, 157.

D

Dairy Foods, 14.
Dairy Industry, 152.

Dairy Products, 3, 20, 40, 44, 46, 56, 99, 122, 152, 206.
Detergent Formulation, 161.
Developments, 60, 62, 81, 82, 83, 94, 96, 97, 117.
Dietary Calcium Supplements, 210.
Dough Conditioning, 136.
Drug Safety Evaluation, 41.

E

Ecological Efficiency, 7, 9, 10, 11.
Emulsion, 21, 143.
Energy Density Concept, 119.
Environmental Biotechnology, 88.
Environmental Impacts, 87.
Enzymes, 59, 60, 61, 62, 63, 64, 65, 66, 67, 68, 69, 70, 71, 72, 73, 74, 75, 76, 79, 80, 81, 82, 84, 86, 87, 88, 89, 102, 103, 104, 120, 135, 136, 137, 138, 139, 140, 141, 142, 143, 144, 145, 146, 147, 148, 149, 150, 152, 153, 154, 155, 156, 158, 159, 160, 161, 162, 163, 164, 180, 181, 185, 186, 207, 235, 237, 238, 239, 240, 241, 242, 243, 254, 259, 266, 267, 268.
Enzymology, 135.

F

Film Boiling, 25, 26.
Food Authority, 36.
Food Chain, 3, 5, 6, 7, 12, 99, 118.
Food Industry, 1, 42, 50, 51, 93, 95, 97, 106, 112, 113, 116, 135, 165.
Food Irradiation, 110, 111, 117.
Food Manufacture, 50, 51.
Food Nutrition, 199.
Food Poisoning, 35, 36, 37, 55.
Food Preservation, 50, 56, 109, 222.
Food Processing, 43, 50, 51, 53, 59, 60, 64, 79, 96, 100, 104, 107, 110, 135, 145, 156.
Food Product, 44, 50, 53, 95, 96, 98, 111, 132, 133.
Food Production, 12, 15, 55, 86, 94, 101, 104, 116.
Food Protection, 1.
Food Safety, 1, 17, 24, 41, 42, 43, 44, 45, 46, 47, 48, 51, 56, 62, 96, 100, 104, 108, 113, 116, 117, 118, 122, 124, 129, 147.
Food Science, 47, 108, 112, 113.
Food Security, 1, 114.
Food Sources, 1, 14, 121.
Food Webs, 3, 4, 5, 7, 12, 13.
Foodborne Pathogens, 86.
Foundation, 90, 206, 272.
Fruit Juices, 64, 181.
Frying, 21, 22, 25, 28, 29, 49, 86, 103.

G

Geochemical Cycling, 203.
Glucose Oxidase, 165.
Glucose Oxidase, 136, 137, 165, 166.
Glucose Oxidase Electrode, 165.
Government Policy, 268.

H

Health Organizations, 272.
Health Risks, 99, 195.

I

Immobilization, 60, 62, 72, 73, 74, 75, 76, 77, 78, 79, 80, 81, 82, 138, 139, 140, 141, 142, 143, 144, 154.
Immobilized Enzymes, 61, 139, 143, 144.
Industrial Biotechnology, 87.
Industrialization, 53.

Index

Industry, 1, 15, 29, 37, 42, 50, 51, 52, 53, 54, 61, 87, 92, 93, 94, 95, 96, 97, 106, 108, 112, 113, 116, 124, 125, 135, 136, 139, 152, 157, 159, 160, 165, 185, 220, 224, 244, 266.
Infant Formula, 44.
Ingredients, 3, 18, 19, 20, 21, 29, 30, 46, 48, 49, 51, 52, 53, 59, 93, 94, 95, 96, 99, 102, 103, 105, 115, 116, 125, 183, 185, 191, 195, 224, 271.
International Biotechnology Policy, 94.
International Trade, 94.

L

Listeria Monocytogenes, 110.

M

Macronutrients, 118, 121.
Mechanism, 114, 134, 149, 161, 181, 183, 207, 231, 254, 259, 264, 265.
Methionine Salvage, 235.
Microorganisms, 27, 62, 64, 86, 88, 89, 109, 110, 111, 112, 115, 116, 135, 136, 145, 148, 149, 150, 152, 153, 158, 159, 162, 181, 196, 244.
Milk Powder, 176.
Milk Processing, 64, 152.
Milk Protein, 135, 152, 176.
Modern Biotechnology, 83, 84, 86, 87, 89, 90, 101, 103, 116, 127, 128, 129, 131.

N

Nucleate Boiling, 25.
Nucleosynthesis, 201.
Nutrition, 13, 14, 45, 57, 58, 66, 90, 93, 95, 102, 106, 112, 113, 114, 116, 118, 120, 125, 181, 185, 199, 200, 206, 214, 227, 239, 268, 269, 270, 271, 272.
Nutritional Assessment Process, 127.
Nutritional Quality, 101, 125, 126.

O

Organic, 4, 7, 15, 31, 38, 40, 57, 71, 72, 74, 87, 94, 97, 98, 99, 100, 106, 107, 136, 139, 143, 155, 184, 212, 219, 238, 240, 241, 242, 243, 244, 245, 246, 247, 249, 250, 251, 252, 253, 254, 255, 256, 260, 261, 262, 263, 265, 266, 268.
Organic Acids, 31, 184.
Organic Farming, 15, 57.
Organizations, 15, 93, 94, 268, 269, 271, 272.

P

PAH World Hypothesis, 265.
Pasteurization, 110, 111, 167, 178.
Pathogens, 26, 27, 36, 44, 86, 110, 111.
Pectic Enzymes, 180, 185, 186.
Physicochemical Properties, 119, 160.
Powdered Milk, 54.
Probiotics, 113.
Production, 1, 8, 9, 10, 11, 12, 15, 19, 29, 37, 44, 51, 54, 55, 60, 61, 62, 64, 71, 72, 73, 74, 75, 76, 78, 79, 80, 84, 85, 86, 87, 94, 100, 101, 103, 104, 107, 113, 116, 125, 131, 135, 136, 142, 143, 144, 145, 146, 150, 151, 153, 154, 155, 157, 158, 160, 161, 162, 163, 165, 166, 167, 170,

171, 173, 183, 184, 186, 188, 189, 190, 193, 201, 204, 205, 218, 220, 221, 223, 224, 229, 253, 254, 260.
Property, 64, 65, 117, 149, 153, 162, 238, 260.
Protection, 1, 34, 38, 39, 76, 84, 93, 95, 232.
Public Health, 42, 52, 94, 228, 268, 269, 270, 271.
Pyramids, 7, 9, 12.

R

Raw Food Preparation, 49.
Raw Milk, 46, 47.
Regulatory Agencies, 36, 41, 95.

S

Sauerkraut, 101.
Security, 1, 114.
Society, 55, 83, 91, 93, 272.
Sodium, 17, 68, 75, 159, 160, 178, 182, 198, 199, 204, 206, 209, 214, 215, 216, 217, 218, 219, 220, 221, 222, 223, 224, 225, 226, 227, 228, 229, 230, 231, 232, 233.
Sodium-Metabolism, 230.
Solar Energy, 3.
Standardization, 36.
Staphylococcus Aureus, 110.
Sweeteners, 16, 103, 145.

T

Technologies, 19, 45, 51, 62, 83, 97, 108, 109, 110, 112, 114, 117, 120, 127, 136.
Tissue Distribution, 230.
Toxicology Assessment, 130.
Traditions, 15, 19, 105, 189.
Transition Boiling, 26.
Treatment, 15, 26, 34, 56, 88, 109, 110, 111, 112, 137, 140, 147, 151, 159, 162, 163, 164, 196, 226, 241.
Trypsin, 75, 130, 135, 156, 157.

V

Vanillin, 31.
Vegetable Processing, 54.
Vinegar Fermentation, 188.
Vitamins, 1, 13, 22, 27, 57, 59, 96, 101, 103, 118, 124, 130, 133, 234, 235, 236, 237, 238, 241, 267, 270.
Voluntary Health Organizations, 272.

W

Water Sterilization, 26.
Wine Production, 167.
Wood Smoke, 31.
World Health Organization, 36, 53, 110, 128.

Y

Yeast, 20, 104, 136, 140, 143, 152, 170, 172, 173, 175, 176, 177, 179, 186, 191, 193, 205, 240, 241, 266.
Yoghurt, 103, 129, 152.

□□□